Citizen's Band Radio Service Manual

By Robert F. Burns & Leo G. Sands

FIRST EDITION

FIRST PRINTING—NOVEMBER 1971

Printed in the United States
of America

International Standard Book No. 0-8306-1581-4

Library of Congress Card Number: 77-170665

Cover photo courtesy of
SBE Division, Linear Systems, Inc.

Contents

Preface

Millions of CB transceivers are in use, and many require periodic repair and preventive maintenance. The CBer (citizens band operator) often has difficulty in finding a service shop capable of or interested in servicing CB transceivers. There is a big opportunity for the electronics technician in the CB radio servicing field.

This manual describes troubleshooting techniques. how to use test equipment and the common failures that occur in CB transceivers. The following information about each circuit is provided:

How it works (or what it does), a brief description of the circuit and its functions.

What can go wrong, a listing of common **defects**, what **symptoms** are caused by these defects, and what **evidence** there is to indicate that this is the likely defect.

Check-out steps, a listing of tests that can be performed for determining that the circuit is operational.

The circuits are typical examples. Space does not permit publication of all actual circuits used in all makes and models of CB transceivers. Individual stage circuits may differ widely in implementation among makes and models, and even among units bearing the same model number but assembled on different production runs. Also, it is **not** intended in all cases that all of the circuits shown are exactly the same as in some particular transceiver. The purpose of showing circuits is to illustrate principles.

While this manual was not originally designed to be a training course, it is in a way. The theory of operation of the basic stages of CB transceivers is explained, along with a listing of **failures** that can occur, the resulting **symptoms**, and how to **verify** whether or not the listed failure exists. Carefully read the introductory information about troubleshooting in general. Then, to diagnose a particular stage, refer to the pages dealing with specific stages.

Schematics shown are **typical** to illustrate troubleshooting principles. Whenever one is available, refer to a schematic of the actual equipment being serviced. **Remember**; when a schematic shows a PNP transistor, bear in mind that when an NPN transistor is used, the DC voltage polarities are reversed (the collector and base of a **PNP** transistor are **negative** with respect to the emitter, and the collector and base of an **NPN** transistor are **positive** with respect to the emitter).

Leo G. Sands & Robert F. Burns

Chapter 1

Transmitter & Receiver Functions

In most CB transceivers, all transmitter RF stages operate at the same frequency—the selected CB channel frequency. The RF signal is generated by the oscillator whose frequency is determined by a crystal. The various channels are selected by switching the appropriate cyrstal into the circuit.

In most tube-type CB transmitters, the oscillator drives the RF power amplifier stage directly. (See Fig. 1-1.) But, transistor-type transmitters generally have a driver amplifier stage between the oscillator and the RF power amplifier. (See Fig. 1-2.) Some transmitters employ a frequency multiplier stage between the oscillator and the RF power amplifier; in such a transmitter, the oscillator operates at a lower frequency than the other stages, which are tuned to a multiple of the oscillator frequency.

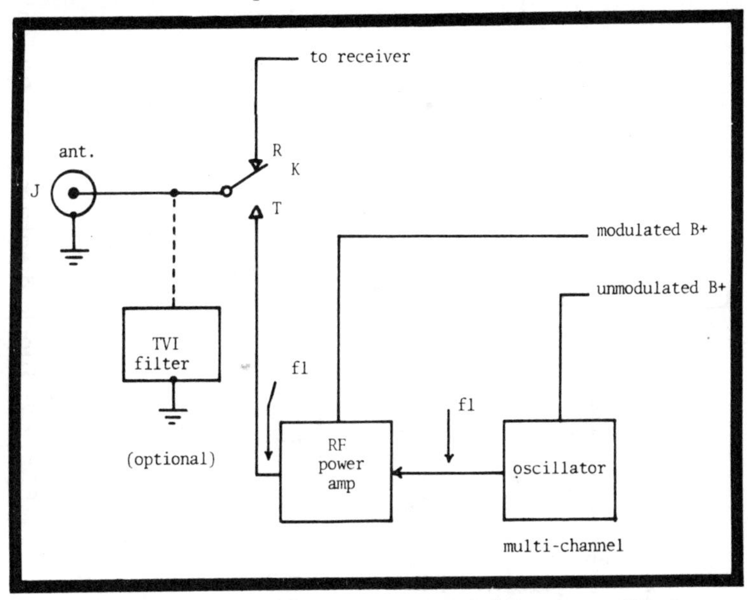

Fig. 1-1. Block diagram of a typical tube-type CB transmitter.

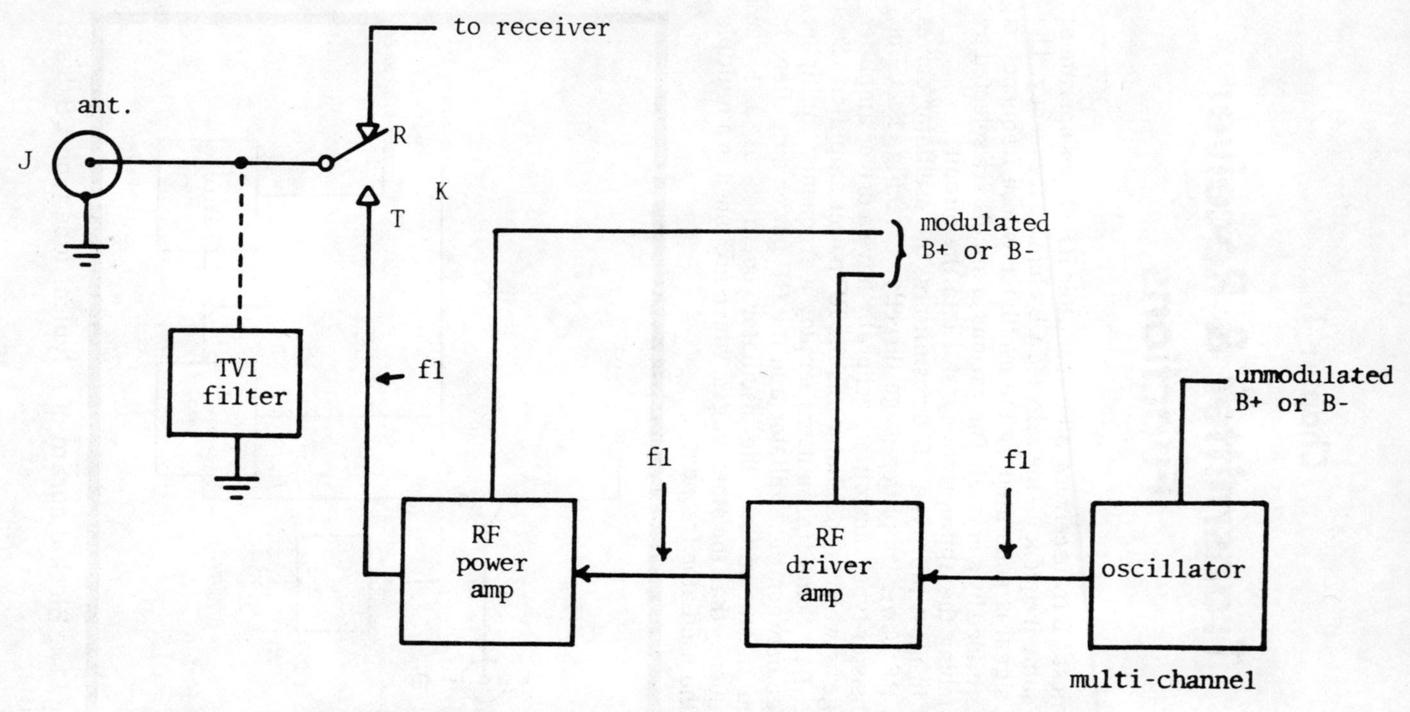

Fig. 1-2. Block diagram of a typical transistorized CB transmitter.

The modulator is an AF power amplifier. It is nearly always used in conjunction with the other AF stages to drive the speaker when receiving and as the modulator when transmitting. The AF output of the microphone is fed to the input of the AF amplifier, directly or through a preamplifier stage. The AF amplifier stages feed the modulator which varies the power output of the RF power amplifier. It does this by feeding AF power in series with the RF power amplifier plate (tube) or collector (transistor) voltage. The AF voltage alternately adds to or subtracts from the DC plate or collector voltage. When 100-percent modulated by a sine wave signal, the RF power output of the transmitter rises 50 percent higher than the unmodulated carrier level. In tube-type transmitters, only the RF power amplifier is modulated. In most transistor-type transmitters, both the driver and RF power amplifier stages are modulated.

Some transmitters employ a modulation limiter in the AF amplifier-modulator circuit. It is sometimes called a "range booster" or "modulation compressor." It prevents overmodulation by clipping or otherwise limiting the level of the AF signal fed to the modulator stage (AF power amplifier). The gain of the AF amplifier is made greater than would otherwise be required. A higher level of modulation is maintained, but overmodulation should not occur since the limiter prevents modulator drive from exceeding a predetermined level.

SUPERHETERODYNE CB RECEIVER OPERATION

All signals (including noise) intercepted by the antenna are fed to the input of the RF amplifier. Its input is broadly tuned so it will pass signals on all 23 channels at the same time. It could be made more selective, but multi-channel operation would not be possible unless elaborate switching or varactor tuning were used.

The RF amplifier provides gain and, in spite of its broad tuning, it also helps reject image interference. The output of the RF amplifier (all CB channels) is fed to the first mixer which also provides some gain. So far, we have not selected a channel. The mixer sees a lot of signals.

To the mixer is also fed a signal from the first local oscillator. It is an unmodulated RF signal whose frequency determines what channel shall be selected. Its frequency is determined by a crystal of which there is one for every channel to be received. One receiving crystal is selected at a time. The oscillator frequency depends upon two things: the channel frequency and the mixer's output frequency, which is known as the "intermediate frequency" (IF).

If the receiver employs a single-conversion superheterodyne circuit, (see Fig. 1-3), there is only one mixer and one local oscillator. If the IF is 1600 kHz, for example, the crystal for Channel 1 would be designed to maintain the oscillator at 25.365 MHz, which is heterodyned with a 28.565-MHz signal, producing 1600-kHz and 55.53-MHz beats. The "sum" beat is rejected by tuning the output of the mixer to 1600 kHz.

Now, we have selected signals on one channel from the many signals present at the input of the mixer. We can select any of the channels by switching in an appropriate crystal. In Fig. 1-3, the incoming CB signal is identified as f1, the local oscillator signal as f2 and the mixer output signal (IF) as f3. The IF signal (f3) always remains the same. It can be produced by making f2 1600 kHz higher or lower than f1. If we change f2, the receiver will respond to an f1 at some other frequency. But, we don't want f3 to change.

The 1600-kHz IF signal is fed to one or more cascaded IF amplifier stages which provide high gain and selectivity. The tuned circuits of the IF amplifier are designed to pass only a narrow band of signals (less than 10 kHz wide). The RF amplifier, on the other hand, is designed to pass a band at least 300 kHz wide.

The amplified IF signal is fed to the detector which is simply a half-wave rectifier. It delivers a DC voltage whose level is proportional to the level of the selected incoming radio signal. This voltage is used to automatically regulate the gain of the RF amplifier, at least one of the IF amplifiers and sometimes the mixer. In a tube-type set this is known as AVC (automatic volume control) and in a solid-state set as AGC (automatic gain control). The purpose of AVC or AGC is to vary receiver gain so that weak signals sound as loud as strong signals.

At the output of the detector, the AF (audio) signal is also developed by rectifier action. The positive (or negative) half of the AM signal envelope is sheared off. If both were present, no AF would result because each half would cancel the other. When one half is removed, the amplitude variations, representing the modulation, become the AF signal.

So far, we have dealt with the incoming CB channel signal (f1) which is an AM signal. The local oscillator signal (f2) is unmodulated. Their heterodyne beat (f3) is an AM signal because f1 is modulated. The modulation extracted at the detector is of varying frequency (f4). The DC extracted at the detector varies in amplitude with the strength of f1 and f3, but not in frequency.

The AF signal (f4) contains the voice intelligence plus all kinds of noise, including noise from auto ignition. The stronger

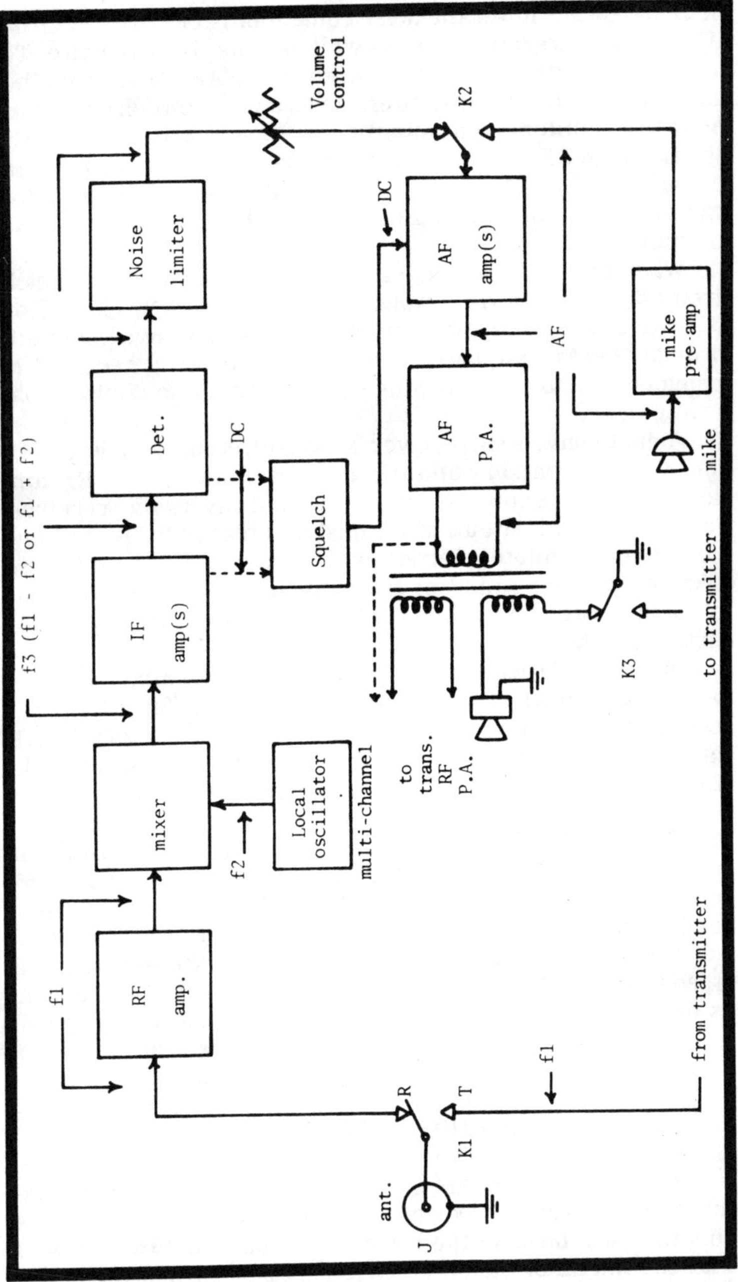

Fig. 1-3. Block diagram of a typical single-conversion CB superhet.

the f1 signal, the lower the noise content of f4 because AVC (or AGC) reduces receiver gain as well as sensitivity to noise. To minimize impulse noise (ignition, etc.), the AF signal from the detector is fed to a noise limiter. When a noise impulse reaches the noise limiter, it automatically blocks passage of AF signals to the AF amplifier. Thus, noise impulses chop holes in the AF signal path. Since noise impulses occur only during a fraction of the time, intelligible AF signals pass through to the AF amplifier.

When no on-channel signal is being received, the squelch circuit disables the AF amplifier. The squelch senses AVC or AGC voltage, directly or indirectly, and opens the signal path through the AF amplifier when the signal being received is strong enough to overcome the squelch. The squelch threshold is adjustable.

A **dual-conversion** receiver is a bit different, (see Fig. 1-4). There are two mixers and two local oscillators. The RF amplifier is the same as in a single-conversion receiver. However, the mixer output is tuned to a higher frequency and the local oscillator accordingly operates at different frequencies. The output of the first mixer is tuned to some frequency between 5 and 10.7 MHz, depending upon the designer's choice.

In a typical receiver, the output of the first mixer is tuned to 6 MHz (600 kHz) and the local oscillator for Channel 1 reception employs a 32.695-MHz crystal. When a 26.965-MHz signal (Channel 1) is received, it is heterodyned with the 32.695-MHz local oscillator signal to produce a 6-MHz "high" IF signal. This signal is fed to a "second" mixer in which the 6-MHz signal is converted to 455 kHz. The second local oscillator operates at 5.555 MHz or 6.445 MHz (6 minus 0.445 equals 5.555 or 6 plus 0.445 equals 6.445). **Note**: Manufacturers employ various "high" IFs; 6-MHz is discussed only as an example.

The 455-kHz or other frequency output of the second mixer is fed to an IF amplifier. The rest of the circuitry is the same as for a single-conversion receiver. Some receivers employ a selectivity filter at the output of the first mixer or the second mixer, depending upon the designer's choice.

SINGLE SIDEBAND OPERATION

When a standard AM signal is used for communications, much of the available power is lost in the carrier. The carrier contains two thirds of the RF power, while the two sidebands that contain all of the audio have only one third of the power. By eliminating the carrier the efficiency of transmission is greatly increased. Then by selecting only one sideband for

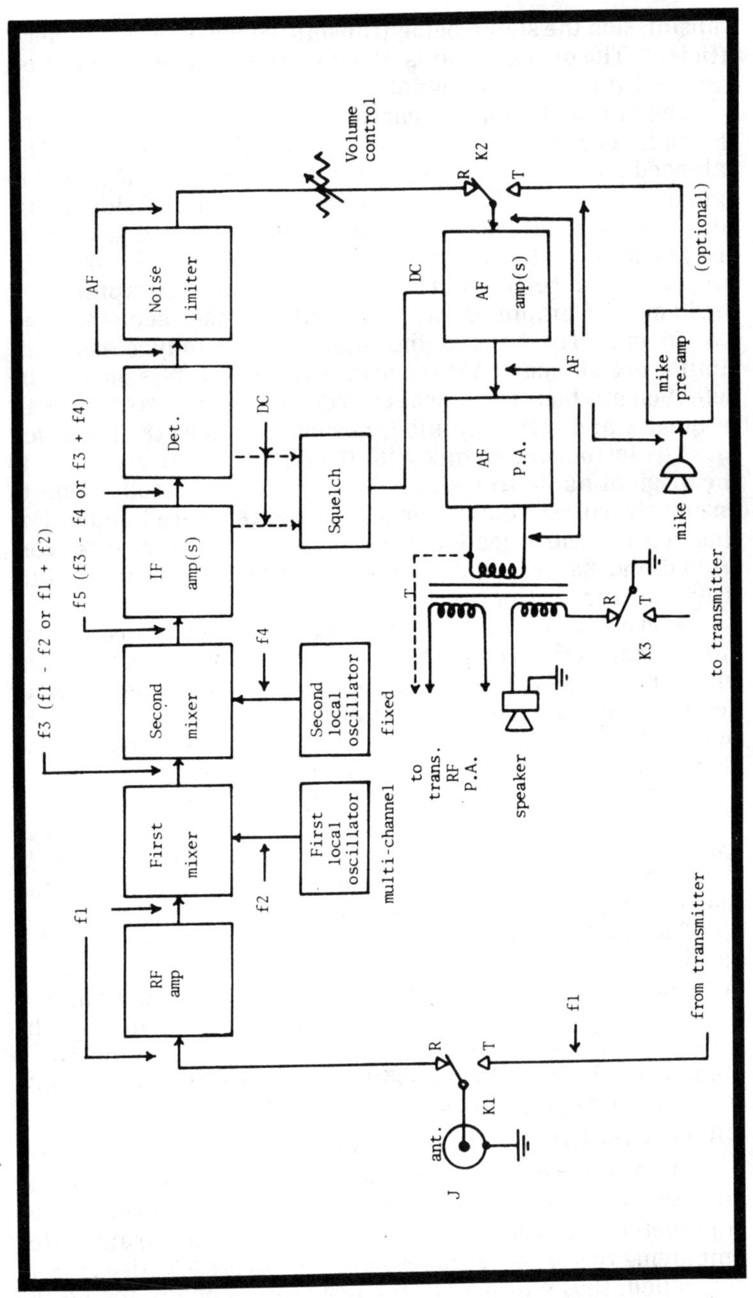

Fig. 1-4. Block diagram of a typical dual-conversion superhet.

transmission the signal being transmitted becomes even more efficient. The overall gain is about eight times what might be expected from an AM signal.

The elimination of the carrier is accomplished by the use of a balanced modulator and an extremely sharp filter. The balanced modulator produces no output until an audio signal is applied; then the "balance" is changed and the sidebands, both sum and difference, are generated. The two sidebands are fed into the filter where one is passed and the other is suppressed or rejected. The single sideband is amplified by the linear RF amplifier and transmitted to the receiving unit.

To properly receive the signal a heterodyne receiver similar to a standard AM receiver is used. It differs only in the detection method. The signal as received is converted to an IF frequency and then amplified; however a fixed RF frequency must be introduced to mix with the SSB signal to reduce it to the original audio frequency. This must be carefully done to ensure the correct quality or tone of the recovered audio. For this reason, the injection frequency is usually crystal controlled and has a slight tuning adjustment for zero beating with the transmitter.

A great advantage is realized with SSB transmission, rather than AM, over paths that are limited by narrow-band interference, selective fading and noise. Since the receiver requires only half of the bandwidth to receive one of the sidebands, the noise and interference are reduced by the narrower bandwidth of the receiver.

When it is necessary to test a SSB transmitter, the use of an oscilloscope is almost a necessity. Another unit that makes testing easy is a low distortion 2-tone generator. The tests should be carried out as described in the instruction manual that accompanies the equipment.

The most common adjustment is the carrier balance. This can be done by using a receiver equipped with an S meter. With no modulation applied to the transmitter, the balance should be adjusted for a minimum reading in the meter. The remaining adjustments on the transmitter are the same as a standard AM transmitter and these adjustments are normally made when the selector switch is in the AM position.

CB TRANSCEIVER SIGNAL FLOW

All AM (amplitude modulated) CB transceivers today employ either a single-conversion or dual-conversion superheterodyne receiver and a crystal controlled transmitter employing two or more stages. Most receivers are also crystal controlled, thus eliminating the need for manually tuning the receiver to the desired channel. The functions of the various stages are expalined in Chart 1-1. (Also see Fig. 1-4.)

Stage	Function	Input signal	Output signal
RF AMPLIFIER (one or more stages)	Gain and image suppression	All CB channels from the antenna	All CB channels to the first mixer
FIRST MIXER	Convert the CB channel frequency to a lower frequency	All CB channels from the RF amplifier plus the unmodulated signal from the first local oscillator	Intermediate frequency (IF) to the IF amplifier directly or through the selectivity filter, in a single-conversion receiver; to the second mixer in a dual-conversion receiver

Chart 1-1. CB transceiver signal flow chart.

Stage	Function	Input Signal	Output Signal
FIRST LOCAL OSCILLATOR	Generate the unmodulated RF signal which is heterodyned with a CB channel signal. Determine the channel frequency by selection of the crystal.	None	Unmodulated RF signal to the first mixer at a frequency equal to the "sum" or "difference" of the CB channel frequency and the first mixer output signal frequency (IF)
HIGH IF SELECTIVITY FILTER (optional)	Readily pass signals within plus or minus 5 kHz or less of the IF center frequency	IF from the first mixer	IF to IF amplifier in a single-conversion receiver or to the second mixer in a dual-conversion receiver
SECOND MIXER (used only in dual-conversion receivers)	Converts the IF signal from the first mixer	IF from the first mixer, directly or	Low IF to the IF amplifier, directly or

	to a still lower IF	through a selectivity filter	through a selectivity filter
SECOND LOCAL OSCILLATOR (used only in dual-conversion receivers)	Generates an unmodulated RF signal which is heterodyned with the IF signal from the first mixer.	None	Unmodulated RF signal to the second mixer at a frequency equal to the "sum" or "difference" of the first mixer output signal frequency and the second mixer output frequency
LOW IF SELECTIVITY FILTER (optional)	Readily pass signals within plus or minus 5 kHz (or less) of IF from the second mixer	IF from the second mixer	IF to the IF amplifier

Chart 1-1. CB transceiver signal flow chart.

Stage	Function	Input Signal	Output Signal
IF AMPLIFIER (one or more stages)	Gain and selectivity	IF signal from the first mixer in a single-conversion receiver or from the second mixer in a dual-conversion receiver, directly or through a selectivity filter	IF signal to the detector
DETECTOR	Demodulate the IF signal; provide AVC or AGC voltage for controlling gain of the RF amplifier, first mixer and-or the first IF amplifier	IF signal from the last IF amplifier stage	AF signal to the noise limiter

NOISE LIMITER	Block passage of noise pulses from the detector to the AF amplifier	AF signal and noise pulses from the detector	AF signal to the AF amplifier
AGC AMPLIFIER (optional)	Increase the dynamic range of AGC (automatic gain control)	Rectified RF signal from the IF amplifier or DC from the detector	DC to RF, mixer and-or the first IF amplifier stages
AF AMPLIFIER (one or more stages)	Gain	AF signal from the noise limiter	AF signal to the AF power amplifier

Chart 1-1. CB transceiver signal flow chart.

Stage	Function	Input Signal	Output Signal
SQUELCH (one or more stages)	Silence the speaker when no signals are received	DC from the detector, the screen of the IF amplifier tube, AVC, or diode which rectifies the IF signal, level of DC varies with the strength of the received signal	DC to the AF amplifier to disable it when no signal is received or activate it when signal is received
AF POWER AMPLIFIER	Power gain and impedance matching to the speaker	AF signal from the AF amplifier	AF signal to the speaker. (Usually also used to modulate the transmitter when transmitting
SPEAKER	Convert electrical waves into sound waves	AF from the AF power amplifier	Sound waves

TRANSMITTER OSCILLATOR	Generate unmodulated RF signal at the CB channel frequency. Determine channel frequency by selection of the crystal.	None	CB channel AF signal to the driver amplifier or RF power amplifier
DRIVER AMPLIFIER (optional)	Power gain	CB channel RF signal from the oscillator	CB channel RF signal to the RF power amplifier
FREQUENCY MULTIPLIER (optional)	Multiply oscillator frequency to the CB channel frequency (seldom used)	RF signal from the oscillator at 1/3 or 1/2 of the CB channel frequency	CB channel RF signal to the RF power amplifier

Chart 1-1. CB transceiver signal flow chart.

Stage	Function	Input Signal	Output Signal
RF POWER AMPLIFIER	Power gain	CB channel RF signal from the oscillator driver amplifier or frequency multiplier	Modulated RF signal at the CB channel frequency to the antenna, directly or through a harmonic filter
HARMONIC FILTER	Attenuate harmonics of the CB channel frequency signal (minimize TVI)	Modulated RF signal from the RF power amplifier	Modulated RF signal to the antenna
MICROPHONE PREAMPLIFIER (optional)	Additional gain	AF signal from the microphone	AF signal to the speech amplifier

SPEECH AMPLIFIER (usually same as receiver AF amplifier)	Gain	AF signal from the microphone, directly or through the preamplifier	AF signal to the modulator, directly or through the roll-off filter and-or modulation limiter
ROLL-OFF FILTER (optional)	Attenuate AF signals above 3000 Hz	AF signal from the speech amplifier	AF signal to the modulator
MODULATION LIMITER (optional)	Prevent overmodulation and maintain high level of modulation	AF signal from the speech amplifier	AF signal to the modulator

Chart 1-1. CB transceiver signal flow chart.

Stage	Function	Input Signal	Output Signal
MODULATOR (usually same as AF power amplifier)	Vary power output of the RF power amplifier	AF signal from the speech amplifier directly or through the roll-off filter and-or the modulation limiter	Variable plate or collector voltage to the RF power amplifier and, in some the transmitter, also to the drive amplifier
VIBRATOR POWER SUPPLY (tube-type sets only)	Convert DC to AC	DC from the battery	AC to the rectifiers; DC from the rectifiers to the tubes
TRANSISTOR SWITCHING POWER SUPPLY (tube-type sets only)	Convert DC to AC	DC from the battery	AC to the rectifiers; DC from the rectifiers to the tubes

VOLTAGE REGULATOR (solid-state sets)	Maintain DC voltage constant	DC from the battery	DC to the transistors
AC POWER SUPPLY	Convert AC to DC. Provide AC when required for the tube heaters	AC from the power line	DC to the tubes or transistors. AC to the tube heaters

Chart 1-1. CB transceiver signal flow chart.

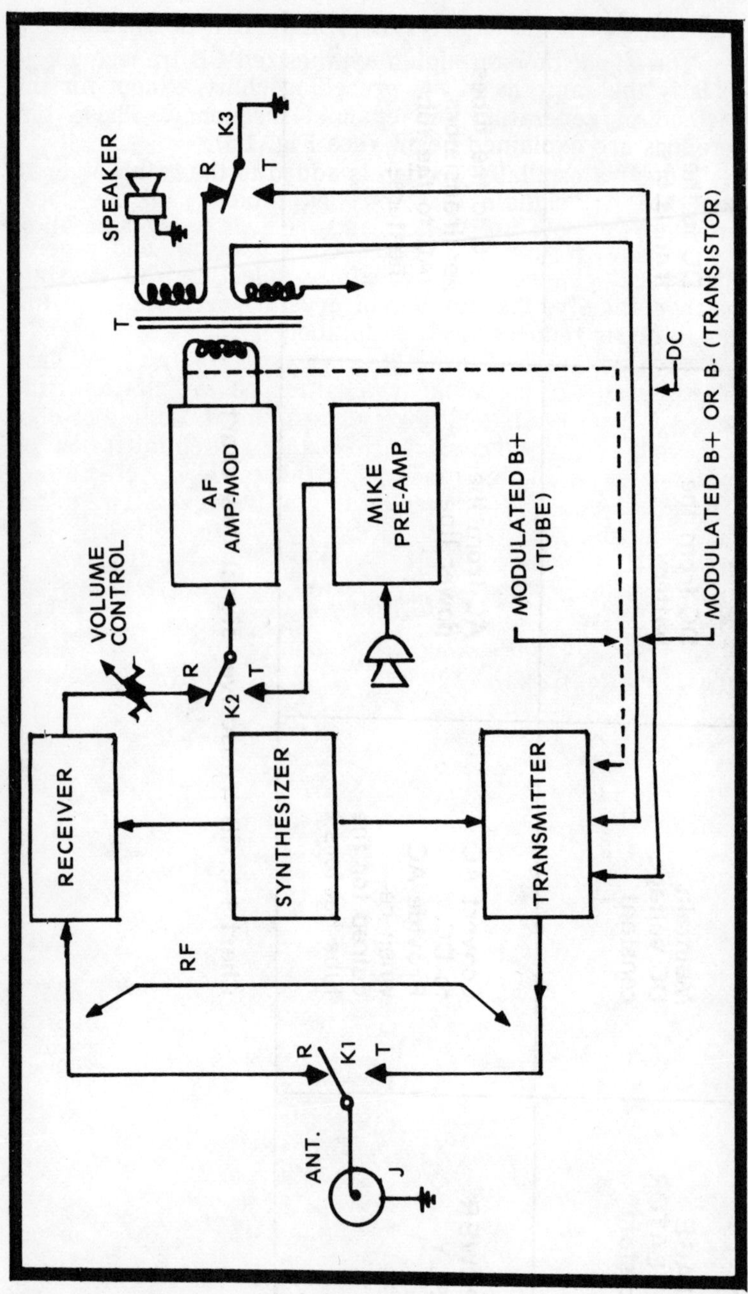

Fig. 1-5. Functional block diagram of a synthesized transceiver.

SIGNAL FLOW IN A SYNTHESIZED TRANSCEIVER

The signal flow through a synthesized CB transceiver is exactly the same as in the preceding chart, except for the method of generating the channel frequency. These differences are explained below (see Fig. 1-5).

An extra oscillator section is added to the transceiver to generate RF signals that may be mixed with the local oscillators of both transmitter and receiver to produce all of the required frequencies necessary to transmit and receive the 23 CB channels. By carefully selecting the crystals mathematically, the number of crystals necessary for full capability is reduced from 46 to about 14.

The synthesizer oscillator uses six crystals that are heterodyned with either the transmitter local oscillator or the receiver local oscillator. Each of these local oscillators contains four crystals. The channel selector switch must now be wired so that 23 combinations of crystals can be selected.

An example of the heterodyning of the crystals required for Channel 1:

Ch 1R: X (synthesizer) + X (receiver) + (low IF)

Ch 1R: 16.965 MHz + 9.545 MHz + 0.455 MHz) = 26.965 MHz

Ch 1T: X (synthesizer) + X (transmitter)

Ch 1T: 16.965 MHz + 10,000 MHz = 26.965 MHz

Chapter 2

CB Transceiver Checkout

To check a mobile unit, connect the battery cable to a metered, variable DC power supply. **Observe polarity!** (Red is the positive lead.) Connect a No. 47 lamp, RF wattmeter, or SWR meter with a 50-ohm load (RF output indicator) to the antenna terminal. (See Fig. 2-1.) Turn the power switch on. Set the input voltage to 13.8 volts DC (12-volt set) or 6.9 volts DC (6-volt set).

If it is a solid-state (all transistor) mobile unit, the input current should rise immediately to less than 1 ampere. Set the squelch to the unsquelched position (noise in speaker) and the volume up. The input current should rise. Operate the press-to-talk button (transmitter on). The input current should rise to 1 ampere or more.

In tube-type mobile units, the input current should be high at first, falling off as the tube heaters warm up, and then rise as the tubes start to draw plate current. The volume level or adjusting the squelch from mute to noise should **not** cause any variation in input current. Operate the press-to-talk button (transmitter on). The input current might rise, but not significantly. Release the PTT button (transmitter off). Lower the input voltage to 11 volts (or 5.5 volts). Current should drop. If it rises sharply, the vibrator or power transistors are not switching. Raise the input voltage to 14.4 volts (or 7.2 volts). The input current should rise also.

Observe the RF output indicator (see Fig. 2-2), operate the PTT button (transmitter on) and vary the input voltage from 11 to 14.4 volts (or 5.5 to 7.2 volts). RF output should increase as the DC input voltage increases. Key the transmitter on and off on each channel for which the unit is equipped. Operate at minimum and maximum input voltages. RF output should be present every time the transmitter is turned on. Set the input voltage to 13.8 volts (or 6.9 volts) and talk into the microphone with the transmitter on. RF output should rise.

For **base station** checkout, connect the power plug to an AC outlet. Connect a No. 47 lamp, RF wattmeter or SWR

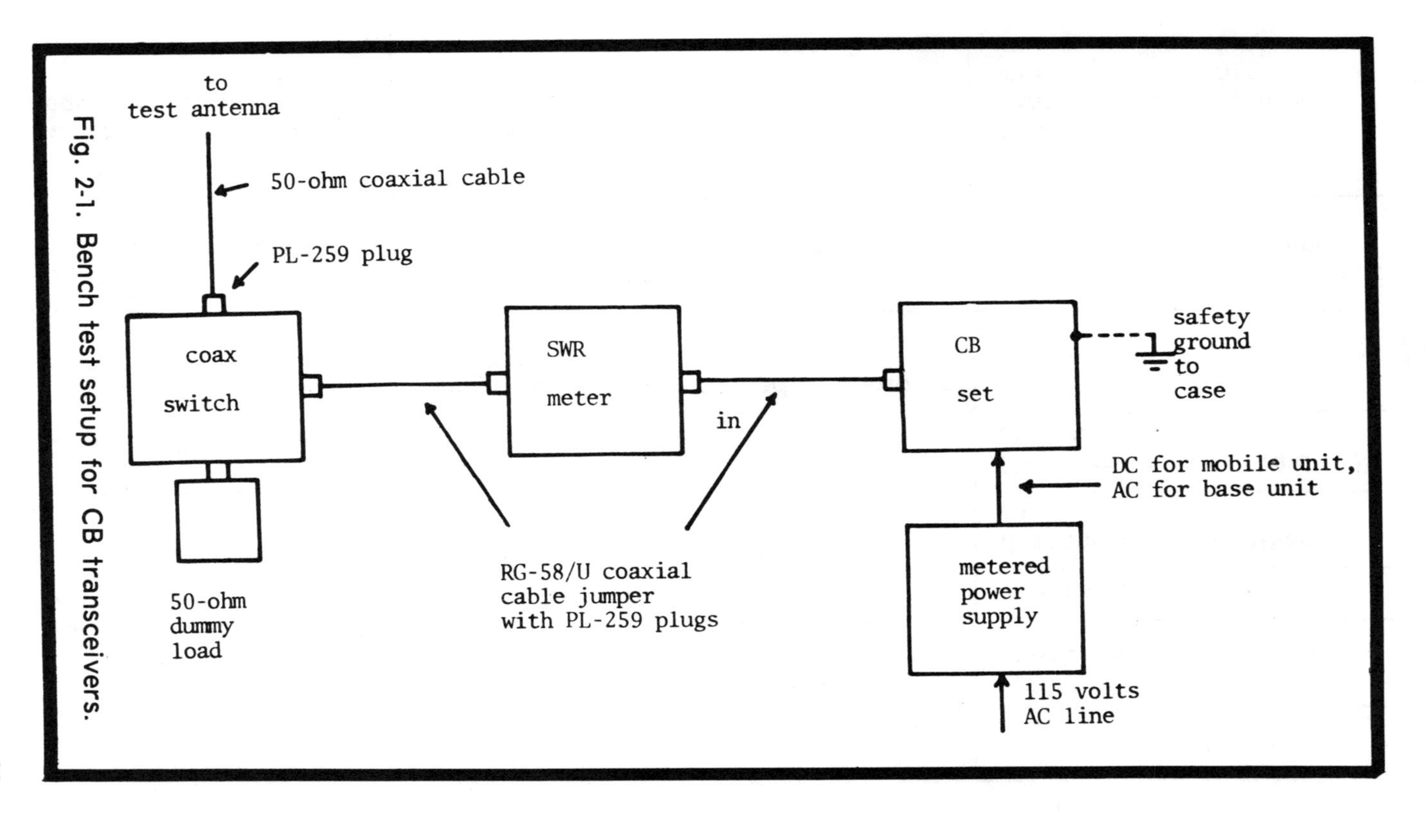

Fig. 2-1. Bench test setup for CB transceivers.

Do This	This Should Happen	If Not, Do This
1. Turn transceiver on, squelch off and volume up.	Pilot lamp lights.	Check fuse, power cord.
	Loud noise in speaker.	See Table 2-4.
	All tubes light.	If some light and others don't, check tubes.
2. Set to transmit.	Speaker is silenced.	Check mike switch and T-R control circuit.
	RF power output indicated by lamp or RF wattmeter.	See Table 2-2.
3. Talk into mike with unit set to transmit.	RF power output rises.	Check mike and connections. Check transmitter tuning.
4. Repeat Step 3 on all equipped channels.	RF power output rises.	If no noise or signals on some channels, try new crystal on dead channels.

5. Connect transceiver to antenna. Set channel selector to all live channel positions.	Loud noise on all channels when no signal being received. Signals received on some channels.	If no noise or signals on some channels, try new crystal on dead channels.
6. Repeat Step 5 after setting squelch to stop noise.	Loud noise not heard on any channel. Signals heard on some channels.	If squelch cannot be set to stop noise, see Table 2-4.

Table 2-1. Basic checkout procedure. Test conditions: transceiver connected to the power source, No. 47 lamp or RF wattmeter connected to the transceiver antenna terminal.

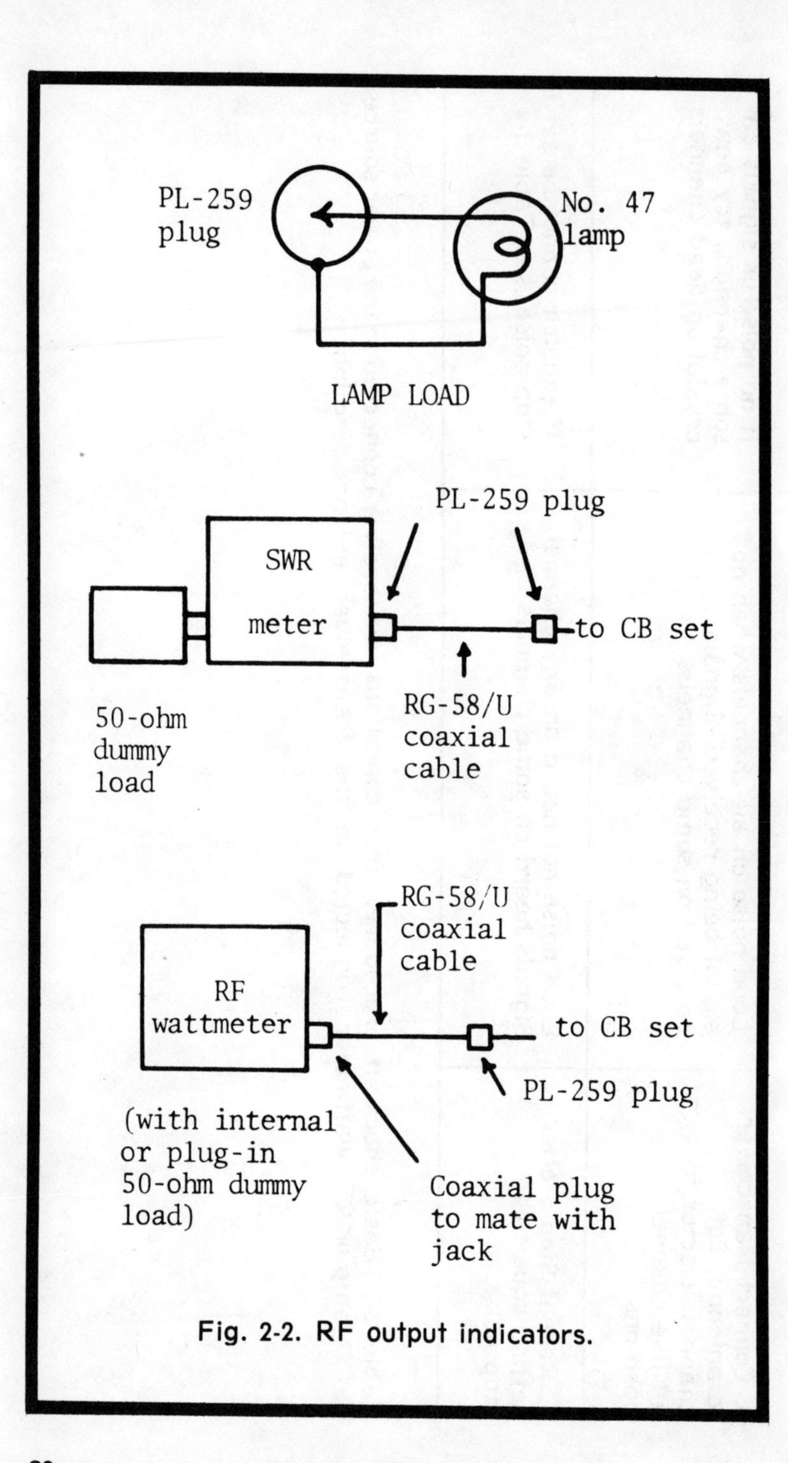

Fig. 2-2. RF output indicators.

meter with a 50-ohm load (RF output indicator) to the antenna terminal. Turn the transceiver on. Adjust the squelch to mute and reactivate the speaker (noise heard). Set to all channels for which the unit is equipped. Noise should be heard on all channels.

Operate the PTT button (transmitter on). Observe the RF output indicator. Key the transmitter on and off on all channels for which the unit is equipped. RF output should be present every time the transmitter is turned on. Repeat while talking into the microphone. The RF output should rise when talking into the microphone. See Table 2-1.

BASIC POWER INPUT TESTS

The quickest way to diagnose significant CB transceiver troubles is to measure the input current with an ammeter. In the case of a mobile unit, connect a DC ammeter in series with one of the battery leads. To measure the input current of a base station unit, connect an AC ammeter in series with one lead of the power cable.

When checking a solid-state transceiver, the ammeter should indicate **low current** with the unit set to receive, the squelch operational (no noise from the speaker), and **no signal** being received. When a signal is received, or with the squelch open (noise in speaker), **current should rise** and increase as the volume is turned up. (True when the AF power amplifier operates Class B.) When the **transmitter** is turned **on, current should rise** to approximately 1 ampere **and** should **vary with modulation** (speaking into mike).

When checking a tube-type transistor, current should be high when it is first turned on and then vary as the tube heaters warm up and plate current rises to normal. Current should **not** vary significantly when a signal is received or squelch is open. (True when the AF amplifier operates Class A.) Current may or may not rise when the transmitter is turned on, depending upon the receiver standby current requirement. **Current should** not vary significantly when the transmitter is **modulated.** See Chart 2-1 and Table 2-2.

BASIC POWER OUTPUT TESTS

Measure transmitter power output with a No. 47 lamp, RF wattmeter (with a built-in 50-ohm load) or SWR meter and 50-ohm dummy load connected to the antenna terminal. Perform the tests in Chart 2-2. See, also, Table 2-2.

	Possible Trouble	
Ammeter Indication	Solid State Unit	Tube-type Unit
Higher than rated current at all times. (Unit does not operate.)	Shorted transistor or capacitor.	Shorted capacitor or stuck vibrator, or shorted power switching transistor.
Lower than rated current at all times. (Unit does not operate.)	Open transistor.	Defective modulator tube or open rectifier.
High current on receive, normal on transmit. (Transmits but won't receive.)	Shorted transistor in the receiver or short in the power distribution network.	Short in the receiver power distribution network.
Higher than rated current on transmit, normal on receive. (Receives but won't transmit.)	Shorted transistor in the transmitter or short in the power distribution network.	Shorted transmitter tube or capacitor, or short in the power dis-tribution network.
Lower than normal current on receive, normal on transmit. (Won't receive.)	Open receiver transistor or resistor.	Open resistor or switching circuit.
Lower than normal current on transmit, normal on receive. (Won't transmit.)	Open transmitter transistor, resistor or coupling capacitor.	Defective RF power amplifier tube.
Current varies at all times (after warm-up, won't operate.)	Leaky capacitor.	Leaky capacitor or worn-out vibrator.
Current very high initially (won't operate), drops later (unit operates).	Marginal capacitor.	Marginal capacitor.
Current normal initially (unit operates), rises later (won't operate).	Marginal capacitor.	Marginal capacitor.
No current at any time (won't operate).	Blown fuse or open main power circuit.	Blown fuse or open main power circuit.

Chart 2-1. Input current checks.

Do This	This Should Happen	If Not, This is the Probable Cause
Turn the transmitter on; **don't talk** into the mike.	RF output steady.	Oscillator not operating, or other defect in the transmitter.
Turn the transmitter on; **talk** into the mike.	RF output rises.	Transmitter improperly tuned, or defect in the modulator, or the mike defective.
Key transmitter **on and off** on all channels.	RF output every time.	Oscillator improperly tuned.

Chart 2-2. Power output tests.

Symptom	Do This	If This Happens	Check This
1. Won't transmit on any channels; receives OK.	Turn the transmitter on with the mike switch.	No indicated RF power output.	Transmitter oscillator and RF amplifiers.
		Same and receiver not silenced.	T-R control circuits, including mike switch.
2. Won't modulate at all.	Same; also talk into mike.	RF output does not change.	Mike and modulator. See Table 2-3.
3. Low modulation.		RF output drops or rises too little.	Transmitter tuning and tubes.
4. Instability, won't transmit every time mike switch is operated.	Test all live channels, keying mike switch several times on each channel.	No RF output every time.	Transmitter oscillator tuning. If on only one channel, try another crystal on that channel.
5. Short range.	Same as for Step 1.	Low indicated RF output.	Transmitter tuning and tubes, power voltages.
6. Hum on carrier	Same as above while listening to carrier with another receiver.	Steady hum when not modulating.	Power supply, modulator and tubes.
	Same as above while talking into mike.	Hum superimposed on voice when modulating, but not on unmodulated carrier.	Modulator.
7. Distorted transmission.	Same as above.	Distorted speech.	Modulation level, modulator, AF tubes.

Table 2-2. Transmitter checkout procedure. Test conditions: Transceiver connected to a power source. No. 47 wattmeter connected to the antenna terminal.

TRANSMITTER TUNING

To tune a CB transmitter, the following instruments are required: RF watt meter with built-in 50-ohm dummy load, SWR meter with external 50-ohm dummy load, or No. 47 lamp (RF output indicator), plus a DC vacuum tube voltmeter with an RF probe.

1. Connect the RF output indicator to the transceiver antenna terminal. (If an SWR meter is used, set it to read "forward" or "incident" power).

2. Turn the power switch on. Set the unit to a mid-band channel. Allow sufficient warm-up time.

3. Turn the transmitter on. Notice the RF output level.

4. Tune the oscillator stage coil for maximum RF output. Then tune the driver or frequency multiplier stage, if either exists. Adjust the RF power amplifier tuning and loading for maximum RF output.

5. Retune the oscillator, noticing that on one side of peak output, power output falls off more slowly than on the other. This is the "gentle" slope. Set the oscillator tuning slightly off peak on the gentle slope. Operate the transmitter on all channels for which equipped, keying the transmitter on and off. Power output should exist every time the transmitter is turned on.

6. Observe the power output on all channels. If it is much higher on some channels than on others, retune for approximately the same output on all channels.

7. Talk into the microphone while transmitting on all channels. Power output should increase when talking into the microphone.

If the transmitter won't tune properly, touch the RF probe of a VTVM to the input of each stage after the oscillator and adjust the previous stage for maximum VTVM indication. However, tune the oscillator slightly below peak output on the gentle slope. If the transmitter has input test points, use the VTVM without the RF probe to measure the DC voltage. Tune the RF power amplifier for maximum output as indicated by the RF output indicator.

8. Place a TV receiver set to Channel 2 near the transmitter. Turn the transmitter on and notice if the TV receiver picture is affected (TVI). If so, tune the transmitter's TVI trap for minimum TVI.

Note: If TVI is severe, disconnect the RF output indicator. Connect the input of the SWR meter to the transmitter antenna terminal through a coaxial cable jumper. Connect the output of the SWR meter to a base station antenna. Adjust the RF power amplifier tuning and loading for maximum indicated "forward" or "incident" power and minimum "reflected" power. Adjust the TVI trap for minimum TVI with the TV set at a distance from the transmitter.

FREQUENCY MEASUREMENT

While the FCC does not stipulate how often CB transmitter frequency measurements should be made, it is nevertheless a good idea to measure the frequencies whenever a transceiver is serviced in a shop. The measurements must be made with a frequency meter accurate to at least 0.0025 percent. The frequency meter should be used as specified in its operating manual. (See Fig. 2-3.)

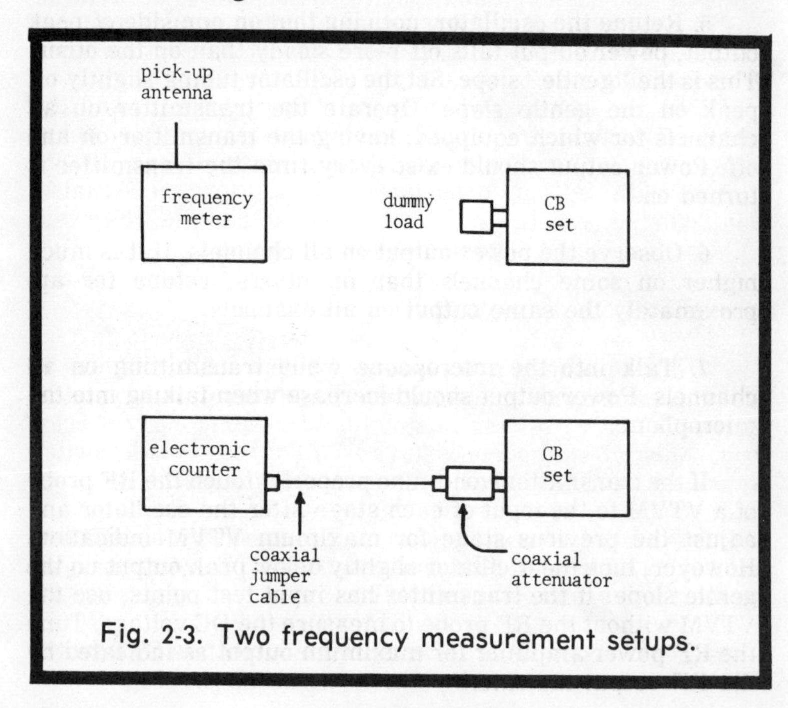

Fig. 2-3. Two frequency measurement setups.

When making frequency measurements, log the actual frequency as measured—not simply as "OK" or "within FCC tolerance." Furthermore, it is a good idea to give the licensee a copy of frequency measurement records. The record should also include: make and model of frequency meter, date of most recent calibration, and signature and operator license of person making the measurement, as well as the date the measurements were made. **Warning:** Off-frequency operation can result in your customer being cited for FCC rule violation.

MODULATION MEASUREMENT

The presence of apparently proper modulation can be determined with an RF output indicator. The RF output should rise with modulation. Maximum talk-out range can be achieved only when the modulation level is high. It should reach close to 100 percent but should never exceed 100 percent on negative peaks (overmodulation).

The modulation level can be observed with accuracy with an oscilloscope connected to the IF amplifier of a CB receiver used for test purposes. As shown in Fig. 2-4, install an SO-239 coaxial connector to the test receiver with its center lead connected through a small capacitor to the plate or collector of an IF amplifier stage. Using a coaxial cable jumper, connect the SO-239 to an adaptor as shown. Connect the vertical input of the oscilloscope to the adaptor.

With the adaptor switch in the "mod" position, and the test receiver set to the same channel as the transmitter being tested, the transmitter's modulation envelope can be observed. With the switch set to the AF position, the signal is demodulated and the modulating signal can be observed. Table 2-3 specifies various modulator checks.

MICROPHONE TROUBLES

A microphone is a **transducer**—it converts variations in sound pressure into an electrical voltage (ceramic, crystal or magnetic) or varies the magnitude of electric current (carbon). Most CB microphones have a PTT (press-to-talk) switch (pushbutton). The following facts are important:

1. **Absence of modulation** can be caused by a defective microphone element, poor contact at the PTT switch, or open or short in the microphone cable or its plug.

2. **No transmission** can be caused by defective PTT switch contacts or an open circuit in the mike cable or its plug.

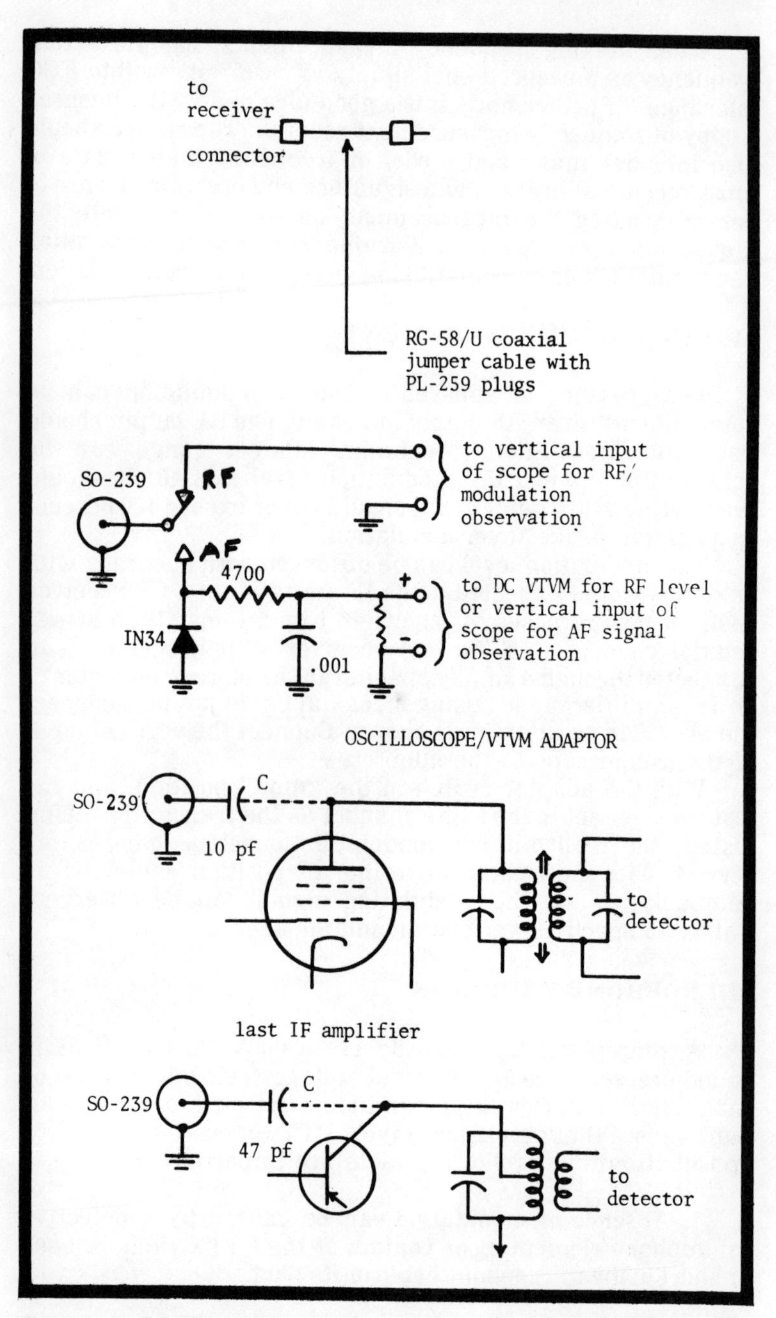

Fig. 2-4. Oscilloscope modulation measurement adaptor and connection diagrams.

Symptom	Do This	If This Happens	Then, Do This
1. No modulation.	Connect AC voltmeter across modulation transformer secondary. Turn transmitter on, talk into mike.	No AC voltage indication when talking into mike.	Try new mike. Apply AF test signal to input of each modulator stage, working back toward mike to identify dead stage.
2. Low modulation.	Same as above.	AC voltage is lower than normal.	Apply AF test signal as above and expect an increase in AC voltage indication as each stage is included in the chain. When no increase, check that stage.
3. Distortion	Connect vertical input of an oscilloscope to the modulation transformer secondary. Apply AF test signal to mike input. Turn transmitter on.	Distorted waveform on oscilloscope.	Connect vertical input of oscilloscope to output of each modulator stage, working back to stage after mike. When clean waveform found, trouble is in stage after it.
	Connect vertical input of an oscilloscope to the primary of the last IF transformer (through a small capacitor). Set receiver to same channel as transmitter under test. Turn on and talk into mike.	Oscilloscope pattern indicates overmodulation.	Reduce gain of modulator by inserting resistance in series with an interstage coupling capacitor.

Table 2-3. Modulator checkout procedure. Test conditions: Transceiver connected to a power source. No. 47 lamp or RF wattmeter connected to the transceiver antenna terminals.

Symptom	Do This	If This Happens	Then, Do This
1. Won't receive on any channel, transmits OK.	Apply modulated RF test signal input of first stage receiver.	Tone modulation heard.	Check antenna relay contacts.
		No tone in speaker.	Check local oscillators. If trouble not there, apply test signal to input of each AF, IF and RF stage, working back from the speaker to identify dead stage.
2. Receives on all channels but one.	Same as above.	No tone in speaker on the dead channel.	Try new receive crystal for that channel. Check channel selector.
3. Receives only strong signals.	Connect AC voltmeter across speaker. Apply RF test signal to input of each IF and RF stage, working back from detector, and watch for increase in voltage as each stage is included in the chain.	No increase in output voltage at particular stage.	Check the stage which provides no gain. If no defect found, realign all stages.

4. Receives but heavy background noise.	Same as for Step 1.	Tone modulation heard in background noise.	Check antenna relay contacts and input coil.
5. Squelch won't unsquelch (continuous noise)	Measure AVC or AGC voltage with VTVM.	Voltage rises when signal received.	Check squelch circuit voltages and tubes.
6. Distorted sound.	Connect headphones to detector output.	Speech is clear.	Check AF amplifier, noise limiter and speaker.
7. Hum in speaker.	Turn volume down to minimum.	Hum still heard.	Check power supply and AF tubes.

Table 2-4. Receiver checkout procedure. Test conditions: Transceiver connected to a power source. No. 47 lamp or wattmeter connected to the antenna terminal.

3. Continuous transmission can be caused by a shorted PTT switch or by a short circuit in the mike cable or its plug.

4. Always replace a microphone with one of the same impedance or output rating.

5. Check the modulation level after replacing a microphone.

BASIC RECEIVER CHECKS

The receivers of CB transceivers are so sensitive that a rushing noise is heard in the speaker when the squelch is open. This noise is the key to receiver testing. Turn the receiver on, the volume up and the squelch open, with or without an antenna connected. Set the channel selector to all channels for which crystals have been installed. **Noise should be heard on all "live" channels.** See Chart 2-3 and Table 2-4.

RECEIVER ALIGNMENT

To align an AM receiver, the following instruments are required: RF signal generator, tunable from 400 kHz to 30

Symptom	Probable Cause
No noise on any channel (transmits OK).	Defect in receiver between detector and antenna.
Noise on all channels but one.	Defective receive crystal for nonoperating channel; or defective channel selector switch.
Noise on some channels, but not all.	Defective crystal when unit employs frequency synthesizer; or defective channel selector switch.
Noise on Channel 1 (or 23) but none on Channel 23 (or 1).	Oscillator improperly tuned.

Chart 2-3. Basic receiver checks.

MHz, and a DC vacuum tube voltmeter. To align a dual-conversion receiver, proceed as follows:

1. Connect the VTVM to the AVC or AGC bus leading from the detector. Connect a No. 47 lamp or 50-ohm dummy load to the antenna terminal. Connect the ungrounded lead of the RF signal generator through a 1000-pf capacitor to the input of the second mixer. Connect the ground lead to the chassis or common ground bus. Remove the second local oscillator crystal. Turn the receiver on and let it warm up for several minutes.

2. With the signal generator modulation on, tune the signal generator to the receiver IF (usually 455 kHz). The modulating tone should be heard in the receiver speaker. Observe the VTVM. Voltage should be present. Lower the signal generator output level so that both noise and the tone are heard.

3. Turn the modulation off. The VTVM should still indicate the presence of AVC or AGC voltage. Tune the signal generator to obtain maximum voltage reading. If the calibration of the signal generator is known to be accurate, tune it to exactly the receiver IF.

4. Starting with the last IF transformer, tune all IF transformers (working back from the detector toward the antenna) for maximum VTVM indication. Lower the signal generator output to a level that still gives a meter reading which varies with the signal generator output level. Repeat the IF transformer tuning. Now tune the signal generator slowly 5 kHz higher than the IF and increase the output level to obtain the same VTVM reading. Repeat with the signal generator set 5 kHz below the IF. The signal generator output level should be the same for the same VTVM readings at both 5 kHz above and below the IF. Otherwise, the IF tuning is not symmetrical.

5. Disconnect the ungrounded RF signal generator lead from the input of the second mixer and connect it to the input of the first mixer. Re-install the second local oscillator crystal. Set the channel selector to an unused channel, or set it to any channel and remove the receiving crystal for that channel.

6. With the output level low, tune the signal generator for the maximum VTVM reading (same frequency as the output of the first mixer). Tune the first mixer output transformer for the maximum VTVM reading.

Defective Stage	Effect	Test Signal	Result if OK
AF power amplifier	Won't receive; won't modulate	AF to input	Tone in speaker
AF amplifier	Won't receive; won't modulate	AF to input	Tone in speaker with squelch off
Mike pre-amplifier	won't modulate	AF to mike input	Tone modulates transmitter
Squelch	won't receive	AF to AF amplifier input	Tone in speaker
Noise limiter	Won't receive	AF to detector output	Tone in speaker
Detector	Won't receive	IF to last IF stage output	Tone in speaker
AGC or AGC amplifier	Distorted reception	RF to antenna terminal	Strong signal quiets back-ground noise
IF amplifier	Won't receive	IF to input	S meter indicates signal. AVC or AGC volt-age rises with signal level.
Selectivity filter	Won't receive	IF to input	Same as above.
Second mixer	Won't receive	IF to input	Same as above
Second local oscillator	Won't receive	RF at crystal frequency to output	If reception restored, crystal or other oscilla-tor defect is apparent
First mixer	Won't receive	RF at input	S meter indi-cates signal; AVC or AGC volt-age rises with signal level
First local oscillator	Won't receive on any channel	RF at crystal frequency to output	If reception is restored, oscillator is not operating

Defective Stage	Effect	Test Signal	Result if OK
	Receives on all channels but one	Same	If reception is restored, crystal or channel switch is defective
RF amplifier	Won't receive	RF at input	S meter indicates signal; AVC or AGC voltage rises with signal.
	Noisy reception	RF at antenna terminal	Signal quiets noise
RF power amplifier	Won't transmit	RF at input	VTVM with RF probe indicates RF present at tank coil
Harmonic filter	Won't transmit	RF at input	Same as above except measure RF at antenna terminal
Driver amplifier	Won't transmit	RF at input	Same as above except at RF power amplifier input
Transmitter oscillator	Won't transmit on any channel	RF at output	Same as above; if RF signal now present, oscillator is defective
	Transmits on all channels but one	RF at output	Same as above; if RF signal now present, crystal or channel switch is defective

Chart 2-4. Effects of stage failures.

7. Disconnect the ungrounded signal generator lead from the input of the first mixer and drape it around the dummy load at the antenna terminal. Reactivate the first local oscillator by re-installing the crystal or setting the channel selector to a live channel.

8. Set the signal generator so it can be tuned to the 26.96-27.26 MHz band. Tune it slowly for the maximum VTVM reading. Keep the signal generator output very low. Tune the first mixer input and RF amplifier coils for the maximum VTVM reading.

9. Set the receiver to the highest channel for which it is equipped and tune the signal generator for the maximum VTVM reading. Then set the receiver to the lowest channel and repeat as above. In both cases, the VTVM reading should be essentially the same. If not, retune the RF amplifier and mixer input coils so that almost the same signal generator output level produces the same VTVM reading on all channels. If not possible, tune the coils for the best performance on the most vital channel without significant degradation on other channels.

To align a single-conversion receiver, perform Step 1, but connect the signal generator to the input of the first mixer (only mixer). Then, perform Steps 2, 3, 4, 7, 8 and 9.

Alignment of Synthesized Transceivers

To align a synthesized transceiver, follow the procedure as outlined previously for the transmitter in Steps 1 through 7 and for Steps 1 through 8 of the receiver section. Then, with the connections the same as in Step 8, tune the output of the synthesizer stage for a maximum VTVM reading. Step 9 should then be done to ensure that the oscillators are working on all channels.

RF Signal Generator Connections

For aligning the receiver RF amplifier and mixer, no direct connection to the RF signal generator is necessary. Simply connect a 50-ohm dummy load to the antenna terminal and drape the signal generator cable near the dummy load. To align the other stages, the RF signal generator output should be connected directly to the stage ahead of the ones to be aligned. Connection can be made through a length of coaxial cable and clips as shown in Fig. 2-5. The clip attached to the capacitor is connected to the signal injection point. The other clip is connected to the chassis or common ground bus.

SQUELCH TROUBLES

The squelch circuit mutes the speaker until the received signal level is greater than the squelch threshold, which is variable. The squelch circuit employs a DC amplifier which is sensitive to minor voltage and resistance variations. The following facts are important:

1. In essentially all CB transceivers the squelch senses the incoming signal level, whether it is a CB signal, diathermy interference or noise.

2. The squelch either controls an AF amplifier stage or acts as a gate in the AF signal patch, depending upon the type of circuit.

3. The squelch may fail to operate or be unstable when a resistor has changed value, a tube in the squelch or controlled AF stage is weak or defective, or when a diode or transistor is open or shorted.

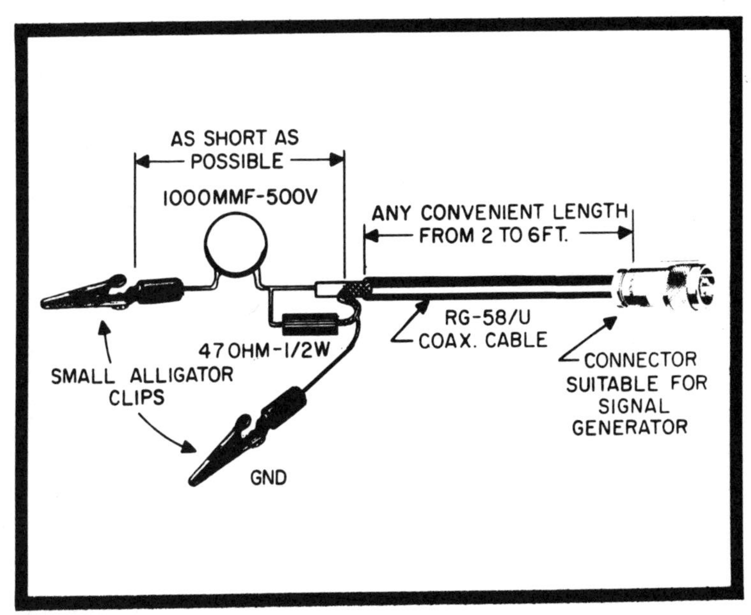

Fig. 2-5. Sketch of the signal generator recommended for RF alignment.

4. Squelch circuit voltages should be checked with a VTVM; a VOM might load the circuit sufficiently to cause the voltage readings to be erroneous.

5. A defective squelch control potentiometer can cause erratic operation or nonoperation.

S METER TROUBLES

An S meter indicates the relative strength of incoming signals in "S" units and db (decibels) above S9. In a tube-type receiver, it usually senses the change in cathode current of an IF amplifier stage in which the gain is controlled by the AVC voltage. In a solid-state receiver the S meter usually monitors the rectified RF level at the detector. The following facts are important:

1. An S meter is not necessarily accurate, and differing S meter readings among receivers do not reflect receiver quality.

2. Often, the S meter is also used to indicate relative transmitter output, in which case the meter indicates rectified RF level.

3. With the transceiver turned off, the S meter can be set to zero with a meter-front screw (if the meter has a mechanical zero adjustment).

4. Many CB transceivers are quipped with internal potentiometers for setting an S meter to zero and to the maximum reading.

5. Always replace an S meter with one of identical electrical characteristics.

SPEAKER TROUBLES

A loudspeaker is a transducer—it converts changes in electric current into changes in air pressure. It essentially always is of the PM (permanent magnet) type with a cone diaphragm attached to a moving voice coil. The following facts are important:

1. No sound from a speaker can be caused by improper contact at the transmit-receive relay, a defective output transformer, or open speaker voice coil.

2. Distorted sound can be caused by the voice coil rubbing against the magnet, or by dirt in the magnet aperture.

3. To determine if a speaker is defective, connect a test speaker to the transceiver speaker terminals.

4. Always replace a speaker with one of the same impedance. The replacement speaker may have a heavier magnet.

5. Don't attempt to repair an inexpensive speaker; replace it.

SIGNAL TRACING

Locating a defective stage is easy when using a multivibrator signal injector (EICO PSI, Don Bosco "Mosquito", etc.) which looks like a fountain pen. Just turn it on and touch its probe to the input of every receiver stage (except local oscillators) starting with the ungrounded speaker terminal. With the volume turned up and the squelch off, a beep should be heard in the speaker each time the probe is touched to the input of a stage. For example, if a beep is heard when touching the probe to the input of the second IF amplifier stage, but none is heard when feeding the test signal to the input of the first IF amplifier stage, it is an indication that the defect is in the first IF amplifier stage.

However, if a beep is not heard when touching the probe to the input of a mixer stage, but is heard when a signal is fed to the next stage, the trouble could be in the mixer stage. Or, the trouble could be in the associated local oscillator stage. In that case, use a tunable RF signal generator. Connect its ungrounded lead through a 10-pf capacitor to the input of the mixer and its ground lead to the chassis. With the RF signal generator modulation off, tune the signal generator to the local oscillator crystal frequency. If the receiver now operates, it is the oscillator stage that is not working.

INTERMITTENT OPERATION

It is often time-consuming and difficult to locate causes of intermittent operation. The receiver or transmitter may function normally and then cut out, only to again function normally before the trouble has been located. A quick way to locate loose connections and other erratic defects in a tube-type receiver is to feed a very strong unmodulated signal directly into the antenna terminal from an RF signal generator. Tune the signal generator to the channel to which the receiver is set. The strong signal will quiet the receiver. Then use the eraser end of a wooden pencil to tap the tubes and prod various components. Trouble spots are identified by noise in the receiver speaker. Using the same technique with a receiver employing transistors might result in damage to the RF amplifier transistor.

Another way to locate causes of intermittent operation is to operate the unit at a higher than normal input voltage to hasten the total failure of a marginal component. Still another way is to raise the temperature of the unit with an infrared lamp or by placing a box over it.

Look for capacitors with loose leads, resistors with cracked surfaces, cold-solder joints, erratic crystals, frayed insulation on wires, dirty switch and relay contacts and tubes with intermittent shorts. Some tube troubles may not become evident until the unit has run for half an hour or longer.

CAPACITOR TROUBLES

Capacitors are used in CB transceivers to resonate coils to a particular frequency, to bypass RF signals to ground, to couple RF and AF signals from one stage to another, to hold a DC charge, and to filter out power supply ripple. A capacitor may be adjustable or fixed. A fixed capacitor may be of the dry or electrolytic type. The following facts are important:

1. A dry (mica, ceramic, paper, etc.) capacitor should not pass DC in either direction after becoming charged.

2. An electrolytic capacitor will pass DC more readily in one direction than in the other.

3. A capacitor may become open, shorted, or leaky.

4. The capacitance of an electrolytic capacitor can be reduced by dehydration and can thus "wear out."

5. Always replace a capacitor with one of the same capacitance and the same or higher working voltage rating.

COIL TROUBLES

Many coils are employed in a CB transceiver—RF, IF and oscillator coils and transformers, RF chokes, AF and power transformers, and filter chokes. Some coils have an air core, some have an iron or steel laminated core. The following facts are important:

1. A coil can be open, have a shorted turn, or be **grounded.**

2. An **IF** or **RF transformer** assembly can contain one or more built-in **capacitors,** any of which can open or become shorted.

3. An IF or RF coil can be checked for inductance with a dip meter.

4. Always replace a coil with an exact duplicate or one with the same electrical characteristics.

5. Always realign a transmitter or receiver after replacing an RF or IF coil.

RESISTOR TROUBLES

Fixed resistors are usually of the carbon composition type. Heavy-duty fixed and tapped resistors are usually wirewound. Variable resistors (rheostats and potentiometers) have a carbon composition element when the power to be handled is low (volume and squelch controls) or a wirewound element when used to control a significant amount of power. The following facts are important:

1. Excessive heat causes the resistance value of carbon composition to change.

2. Charring or discoloration of a resistor is evidence of overheating.

3. A noisy volume control can be cleaned with an electric contact cleaning solvent. If cleaning does not eliminate the trouble, the potentiometer should be replaced.

4. Always replace a resistor with one of the same ohmic value and the same or higher power rating.

5. Do not use a wirewound resistor to replace a carbon composition resistor.

Application	Defect	Symptom(s)
Filter	Dehydrated	Hum, motorboating
RF coupling	Open	Insensitivity, no operation
AF coupling	Open	Low volume, no sound
AF coupling	Shorted, leaky	Distortion
Bypass:		
- plate supply	Open	Insensitivity, oscillation
- plate supply	Shorted	No operation
- collector supply	Open	Insensitivity, oscillation
- collector supply	Shorted	No operation
- cathode resistor	Open	Reduced gain
- cathode resistor	Shorted	Distortion
- screen grid	Open	Insensitivity, oscillation
- screen grid	Shorted	No operation
Resonator (across coil)	Open	Insensitivity
Resonator (across coil)	Shorted	No operation
Neutralizer	Open	Oscillation
Neutralizer	Shorted	No operation
Feedback (oscillator)	Open	No operation
Feedback (oscillator)	Shorted	No operation

Chart 2-5. Capacitor-caused defects.

Symptom	Do This	If This Happens	Then, Do This
1. Transceiver inoperative.	Replace fuse.	New fuse blows.	Look for shorted capacitor, stuck vibrator or grounded circuit.
2. Transceiver inoperative, fuse OK.	Check rectifiers.	Rectifiers OK.	Try new vibrator or check power transistors, power transformer and resistors in filter circuit.
3. Receiver OK, transmitter doesn't operate.	Check transmitter voltages.	No transmitter voltage.	Check T-R relay and resistors in filter circuit.
4. Transmits but won't receive.	Check receiver voltages.	No receiver voltage.	Same as above. Also check voltage regulator if there is one.
5. Operates but gives off odor.	Measure input current.	Input current excessive.	Suspect power transformer which may run too hot.

Table 2-5. Power supply checkout procedure. Test conditions: Transceiver connected to a power source. No. 47 lamp or RF wattmeter connected to the transceiver antenna terminal.

CHANNEL SELECTOR TROUBLES

A channel selector is usually a multiposition rotary switch. Its contacts can become dirty or worn and fail to make positive contact. The following facts are important:

1. Contacts should **not** be bent or cleaned with an abrasive.

2. Switch contacts may be cleaned with an electric contact cleaning fluid.

3. Always replace a channel switch with a mechanical and electrical equivalent.

4. **Do not** attempt to repair a channel selector switch—replace it.

5. When crystals are soldered to a switch, and it is necessary to replace the switch, place the crystals in the same position as they were originally, and keep the lead lengths the same in order to avoid affecting the frequency.

CRYSTAL TROUBLES

A crystal is like a tuning fork. It is resonant at a certain frequency, but its resonant frequency is affected by temperature. In a CB transceiver, crystals are used for determining transmitting and receiving frequencies. The following facts are important:

1. Always replace a crystal designed to operate in the specified make and model transceiver.

2. When adding channels, order crystals for that specific make and model of transceiver. Never install a general purpose crystal.

3. The operating frequency of a crystal is affected by circuit inductance and capacitance, as well as temperature.

4. Although a crystal was designed for use at a specific frequency in a designated make and model of transceiver, it can operate at the wrong frequency because of a mechanical defect.

5. Always measure the transmitter frequency after installing a new transmitter crystal. If you don't have a

frequency meter, have the frequency measured by someone who does.

VIBRATOR TROUBLES

Very few new CB transceivers employ a vibrator, but many older sets do. A vibrator converts DC into AC. It is an electromechanical device with moving electrical contacts. The following facts are important:

1. Vibrators wear out and require periodic replacement.

2. Vibrator contacts can stick, spark or fail to pass current.

3. A solid-state replacement for a vibrator can cause the power input to be raised higher than the lawful 5-watt limit.

4. Always replace a vibrator with an exact electrical equivalent.

5. When replacing a vibrator, it is a good idea to also replace the buffer capacitor.

6. A mobile transceiver should be turned off when starting the vehicle's engine.

RELAY TROUBLES

A relay is an electromechanical switch that can be remotely controlled. When its electromagnet is energized, its armature is pulled in. When not energized, its armature releases or drops out. The armature moves the switching contacts which mate with stationary contacts. The following facts are important:

1. Relay contacts may be cleaned with a burnishing tool—never with a file or sandpaper.

2. Relay springs and other mechanical features are factory-adjusted and should not be readjusted in the field, except by an expert.

3. A relay requires higher voltage (or current) to pull in than to keep it energized.

4. Always replace a relay with an exact electrical equivalent.

5. Instead of trying to repair a relay, replace it.

TUBE TROUBLES

Standard tubes have a life expectancy of 1000 hours of operation—when operating voltages are not exceeded, and when not subjected to shock and vibration as is common in mobile applications. The following facts are important:

1. New (unaged) tubes might fail much sooner than old tubes which have outlived the "shakeout" period.

2. One weak tube can drastically reduce receiver sensitivity or transmitter output.

3. Tubes, other than rectifiers, should be checked with a tube tester that measures transconductance or dynamic mutual conductance. Testers that measure emission only are not critical enough.

4. Pentode tubes used in RF and IF stages should be checked for grid emission with a grid circuit tester or tube checker with grid emission testing capability when available.

5. Always replace tubes only with one of the same type number or electrical equivalent.

INTEGRATED CIRCUIT TROUBLES

An integrated circuit (IC) may contain a number of transistors in addition to other circuit elements such as resistors, capacitors and diodes. **No attempt should be made to repair an IC. It is a component in itself.**

An IC can develop an open or short circuit. It has one or more inputs and outputs. Furthermore, when shorted, an IC can draw excessive current. To check an IC, measure the input and output signal levels. If not normal, replace the IC with an exact equivalent. Do not substitute another type of IC.

TRANSISTOR TROUBLES

Unlike tubes, transistors do not wear out gradually. They are either OK or no good. In most cases, transistors are soldered in place and cannot be easily removed for testing. However, transistors can be tested with an in-circuit type checker, without removing them from the transceiver (with the transceiver turned off). They can also be checked with a VOM or DC voltmeter, with the transceiver turned on, as explained elsewhere in this manual. The following facts are important:

1. Transistors can be destroyed by heat from a soldering iron, or because of the failure of another component.

2. Transistors do not wear out, but they are subject to catastrophic failure.

3. Always replace a transistor with one of the same type number or electrical equivalent.

4. Do not remove a transistor for testing. Instead, test it in its circuit.

5. To diagnose transistor circuits, use signal tracing techniques.

The receiver input transistor (RF amplifier) is vulnerable to destruction by high-voltage static charges. Touching the antenna can cause a static charge of up to 25,000 volts to reach this transistor. To protect this transistor, a small neon lamp can be connected across the input RF transformer or coil. The neon lamp will fire and short circuit high-voltage charges.

When working on transistor circuits, it is a good idea to discharge the static charge built up on your body by touching the transceiver chassis before touching a transistor lead with a metal tool or probe.

TRANSISTOR VOLTAGES AND POLARITIES

PNP transistors. The collector (C) is negative with respect to the emitter (E) and the base (B) must be negative with respect to E for the transistor to conduct. In circuit 1 in Fig. 2-6, the positive side of the source voltage is grounded. In 2, the negative side is grounded. Yet C is still negative with respect to E. R2 is the output load and R1 is the series emitter resistor, which is not always used. The base is usually connected through a voltage divider to a negative DC source (not shown).

NPN transistors. The collector (C) is positive with respect to the emitter (E) and the base (B) must be positive with respect to E for the transistor to conduct. Circuits 3 and 4 in Fig. 2-6 are identical to 1 and 2 except for polarity. The base is usually connected through a voltage divider to a positive DC source (now shown). See Chart 2-6.

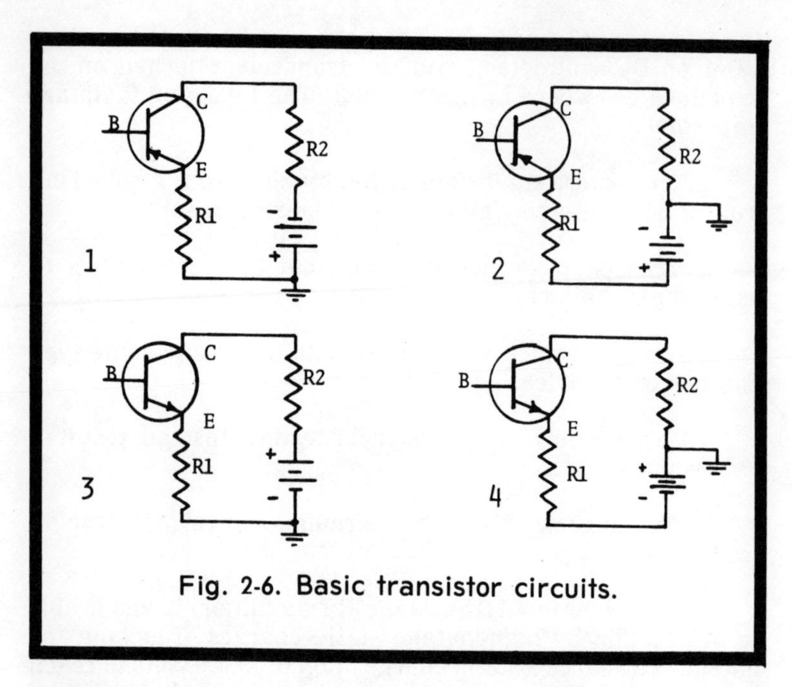

Fig. 2-6. Basic transistor circuits.

Checkout Steps

1. Measure the DC voltage across C-E. Note the meter reading. Short B to E and again note the meter reading. It should rise.

2. Measure the DC voltage across R1. Note the meter reading. Short B to E and again note the meter reading. It should now be lower.

Defect	Symptom	Evidence
Shorted transistor	No operation	No DC voltage across C-E
Open transistor	No operation; no current through R1 and R2	No DC voltage across R1; it will be zero if transistor is open
R1 open	No operation	No DC voltage across R2 except when R1 shorted
R2 open	No operation	No DC voltage across R1 except when R2 shorted with 2200-ohm resistor

Chart 2-6. Transistor trouble symptoms.

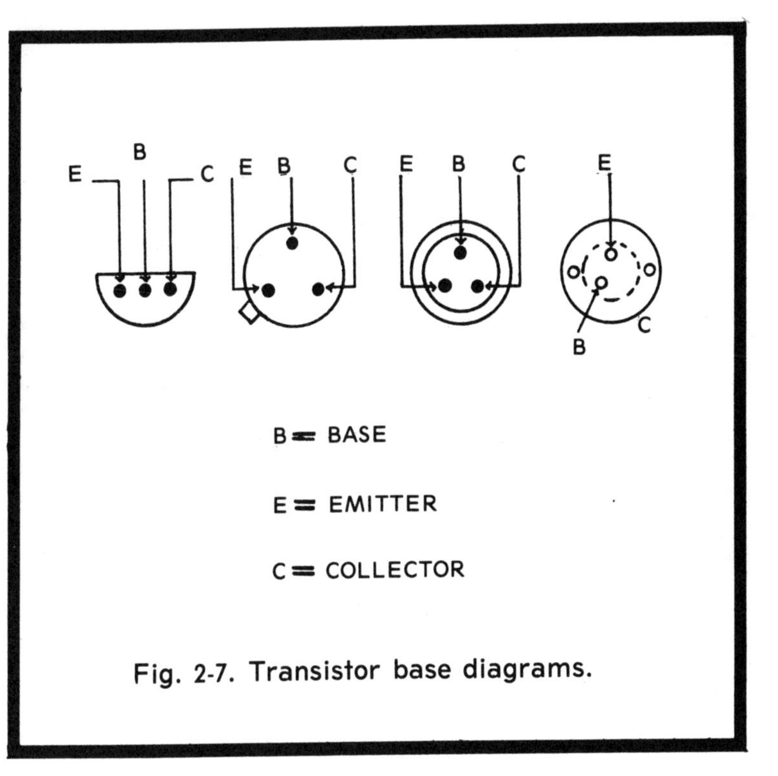

Fig. 2-7. Transistor base diagrams.

Transistor Lead Diagrams

The four illustrations in Fig. 2-7 show the lead locations of various types of transistor packages. Connections to most transistors are made to wires which are soldered to the circuit. The base shown at the far right is of a power transistor which is bolted or riveted to a metallic mass, known as a heat sink, which dissipates heat generated in the transistor. It has soldering lugs instead of leads.

The locations of the leads with respect to each other are shown. The leads are identified by a letter. E is the emitter lead, B is the base lead and C is the collector lead. In the case of the power transistor at the far right, connection to the collector is made through the metal shell, usually at a mounting bolt or rivet.

Connections to PNP and NPN transistors are the same. However, voltage polarities are reversed. The collector and base of a PNP transistor are negative with respect to the emitter. The collector and base of an NPN transistor are positive with respect to the emitter.

Chapter 3

Troubleshooting Transmitter Stages

Transmitter troubleshooting should always begin at the output stage or antenna relay. Usually by proceeding from the final stage back toward the oscillator, the trouble can be found more rapidly. Most problems arise in the higher power sections rather than in the oscillator circuits. Modulator troubles are easily diagnosed by using an oscilloscope. A lack of signal, low amplification or distortion are seen quite readily on the scope. Again the logical method is to apply a tone to the input circuit of the modulator and quickly check from the output back toward the input for a signal. In this way, the problem stage may be quickly isolated and the faulty component replaced or the trouble corrected.

TRANSMITTER OSCILLATOR (Triode)

This oscillator generates the transmitter signal. Switch S in Fig. 3-1 selects the channel crystals (only one crystal

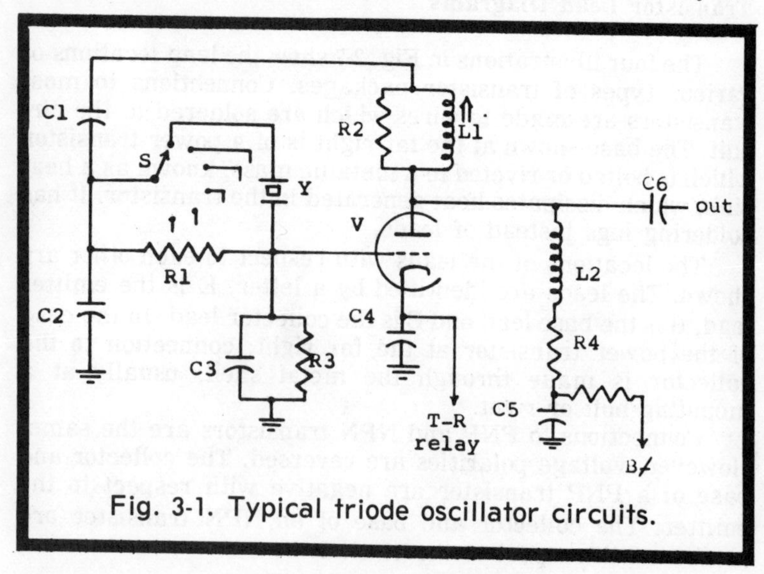

Fig. 3-1. Typical triode oscillator circuits.

Defect	Symptom	Evidence
Weak tube	Won't transmit	New tube restores operation
Bad crystal	One channel out	New crystal restores operation
C1 shorted	Won't transmit	Positive DC voltage across C2
C5 shorted	Won't transmit	No DC voltage across C5
L2 open	Won't transmit	No DC voltage across R4
C4 shorted	Won't stop transmitting	No DC voltage across C4

Chart 3-1. Oscillator troubleshooting hints.

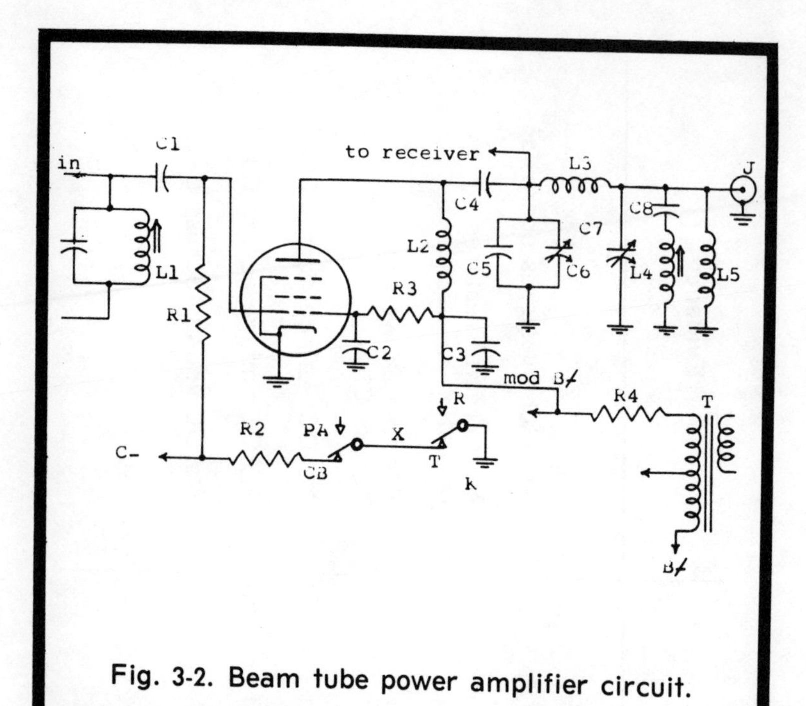

Fig. 3-2. Beam tube power amplifier circuit.

Defect	Symptom	Evidence
C4 shorted	Won't transmit	No DC voltage across C4
C3 shorted	Won't transmit	No DC voltage across C3
C2 shorted	Won't transmit	No DC voltage across C2
R3 open	Won't transmit	No DC voltage across C2
R2 open	Won't transmit	No DC voltage between X and ground when receiving

Chart 3-2. Beam tube RF power amplifier troubleshooting hints.

shown). The oscillator does not operate until the cathode of the tube is grounded by the T-R relay (not shown).

Checkout Steps

1. Measure the DC voltage across R3 through a low-capacity probe. Grid end should be negative.

2. Measure the RF voltage between C6 and ground with a VTVM equipped with an RF probe through a low-capacity probe. Tune L1 for maximum voltage. Then adjust L1 for a lower voltage on the gentle slope of the tuning curve. Key the transmitter on and off several times on each channel. The RF voltage should be present every time.

3. Measure the DC voltage across C4. Voltage should be present in the receive condition, none in transmit condition.

BEAM TUBE RF POWER AMPLIFIER

Signal from the oscillator (Fig. 3-2) is fed through C1 to the control grid of the tube. Its output is fed through a pi-network (C5-C6-C7-L3) to antenna connector J. When switch S is in the CB position and the T-R relay (K) is in the receive position, high negative bias prevents the tube from operating. When K is set to transmit, this bias is reduced by shunting R2 across the C-lead. In the PA position of S, bias cuts off the tube to prevent transmission.

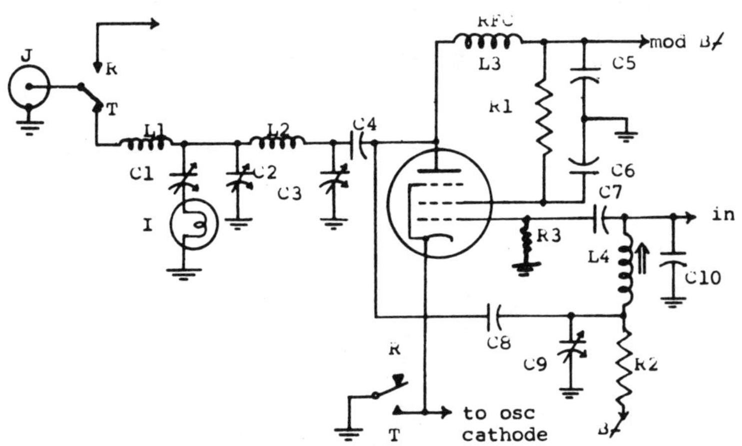

Fig. 3-3. Schematic of a typical neutralized RF power amplifier stage.

Defect	Symptom	Evidence
Weak tube	Low output	Lamp I glows dimly
C4 shorted	Won't transmit	DC voltage present across C3
C4 open	Won't transmit	Lamp I doesn't glow
C5 shorted	Won't transmit	No DC voltage across C5
C6 shorted	Won't transmit	No DC voltage across C6
C7 shorted	Won't transmit	High DC voltage across R3 (grid positive)
C7 open or C10 shorted	Won't transmit	No DC voltage across R3 (grid negative)

Chart 3-3. Troubleshooting data for a neutralized RF power amplifier.

Checkout Steps

1. Connect a No. 47 lamp or RF wattmeter to J. Turn the transmitter on. The power output should be indicated.

2. Same as above. Talk into mike. Power output should rise.

3. Measure the DC voltage across R2. It should be higher when receiving than when transmitting.

NEUTRALIZED RF POWER AMPLIFIER

Signal from the oscillator (Fig. 3-3) is fed through C7 to the control grid of the tube. Its output is fed to antenna connector J through C7, a pi network (C2-C3-L2), L1 and the T-R relay contacts. C9 is a neutralizing capacitor. Lamp I is a power output indicator.

Checkout Steps

1. Connect a dummy load to J. Turn the transmitter on. I should glow.

2. Same as above. Talk into the mike. I should glow more brightly.

3. Disconnect one heater lead at the tube socket so the tube won't light. Measure the RF voltage at J with a VTVM equipped with an RF probe. The voltage should be near zero with transmitter on. If not, adjust C9 for minimum RF output (neutralizing).

PUSH-PULL TRIODE MODULATOR-AF POWER AMPLIFIER

Two triode tubes function as a push-pull AF power amplifier and modulator in Fig. 3-4. AF signals are fed to the grids of V1 and V2 through interstage transformer T1. When the grid of V1 is fed a positive-going signal, its plate current rises. At the same time the grid of V2 is fed a negative-going signal and its plate current decreases. When the grid of V1 sees a negative-going signal, just the opposite is true. The resulting AF voltage developed across the primary of T2 is stepped down and fed to the speaker. The other secondary of T2 steps up the AF voltage which alternately bucks and boosts the B+ voltage fed to the modulated RF power amplifier.

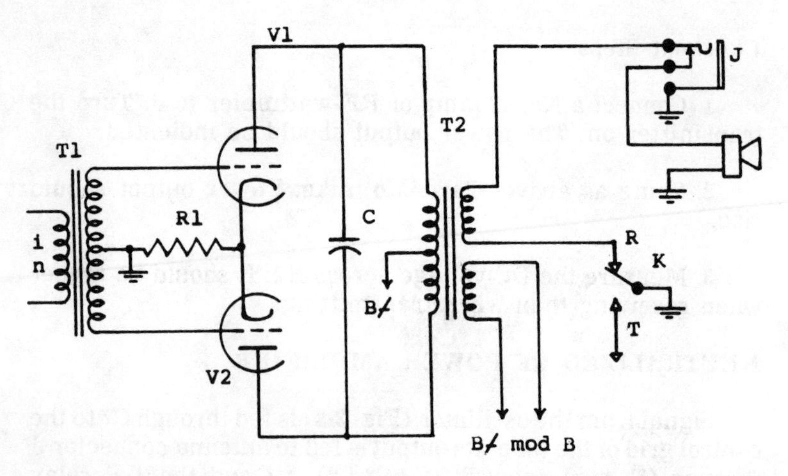

Fig. 3-4. Push-pull triode modulator-AF power amplifier circuit.

Cathode bias is developed across R, and C attenuates the high audio frequencies.

Checkout Steps

1. Measure the V1 plate-to-cathode DC voltage. Shunt R1 with a resistor of the same value. The voltage should drop. Repeat, measuring the V2 plate-to-cathode voltage.

2. Apply an AF test signal to the primary of T1. Measure the AF voltage across R. It should be zero. The AF tone should be audible in the speaker. Check the AF voltage across the modulation secondary of T2.

TUBE-TYPE MODULATOR WITH AUDIO COMPRESSION

V1 in Fig. 3-5 is a squelch-controlled (when receiving) AF amplifier; V2 is the second AF amplifier, and V3 is a beam tube AF power amplifier-modulator. The tapped primary of output-modulation transformer T functions as an autotransformer modulation reactor when transmitting. AF from T is fed through C3 and R7 to diode CR which delivers a negative DC voltage to the grid of V1. When the AF level increases, this negative DC voltage rises, reducing the gain of V1. This provides audio compression and a higher modulation capability without over-modulation. The modulation level is adjusted with R5.

Defect	Symptom	Evidence
V1 and V2 unbalanced	Distortion	DC voltage present across C
R open	Won't operate	Full DC supply voltage across R
C shorted	Won't operate	No AF voltage across C
T1 primary open	Won't operate	No AF voltage across secondary of T1
T2 primary grounded	Won't operate	No DC voltage at plate of V1, V2

Chart 3-4. Troubleshooting tips for a typical push-pull triode modulator-AF power amplifier.

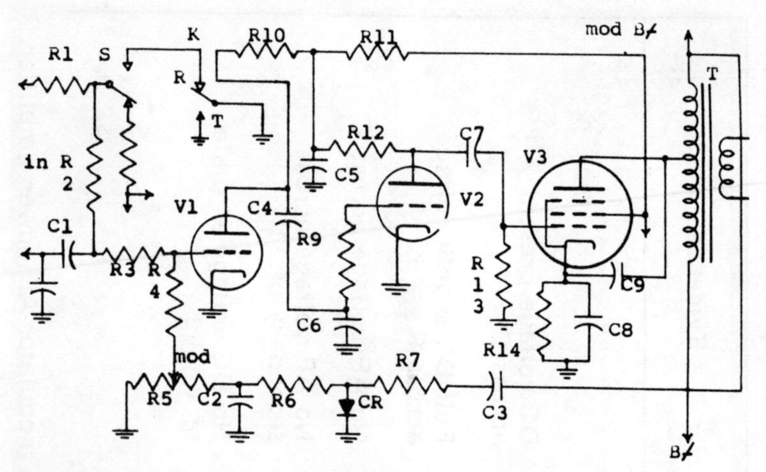

Fig. 3-5. Tube-type modulator circuit with audio compression.

Checkout Steps

1. Apply an AF test signal to C1. Measure the DC voltage across C2. The voltage should rise as the AF test signal level increases.

2. Measure the AF voltage across the secondary of T with an AF test signal applied to C1. Voltage should remain relatively steady, as the AF test signal level is varied.

Defect	Symptom	Evidence
Primary of T open or C9 shorted	Won't modulate-receive	No plate voltage at V3
CR open or shorted	Low modulation	No DC voltage across C2
C5 shorted	Won't modulate-receive	No DC voltage across C5

Chart 3-5. Modulator troubleshooting data.

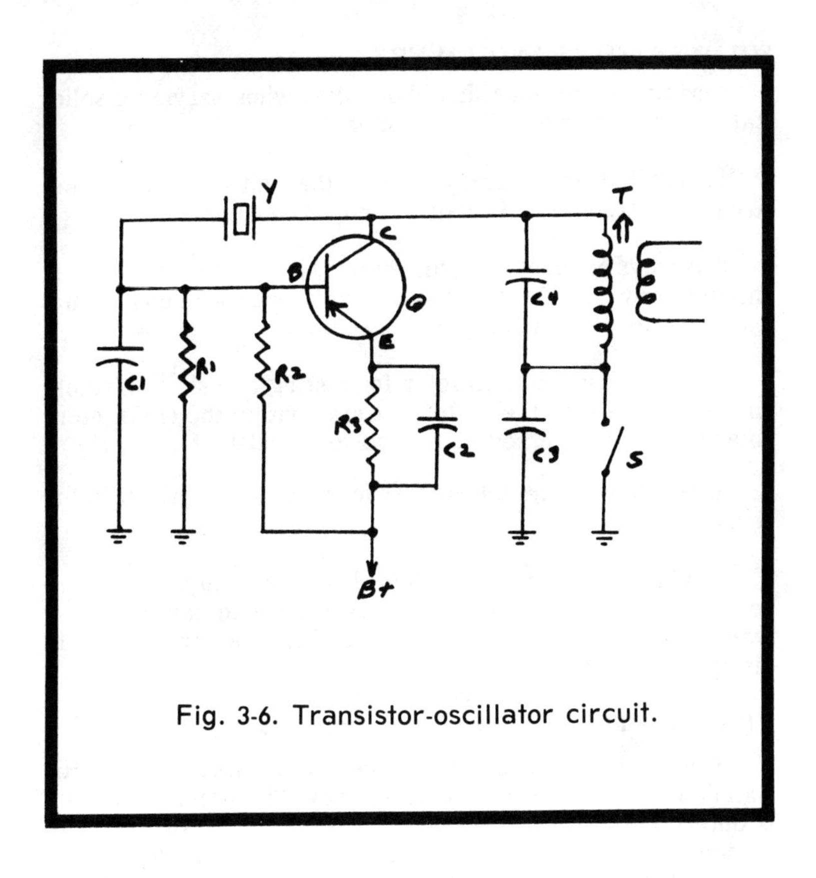

Fig. 3-6. Transistor-oscillator circuit.

Defect	Symptom	Evidence
Shorted transistor	Won't transmit	No DC voltage across C and E of transistor Q with unit in transmit position (S closed).
Defective crystal	Won't transmit	New crystal restores operation
T mistuned	Unstable	Tuning T restores operation

Chart 3-6. Troubleshooting data for transistorized oscillators.

SOLID-STATE TRANSCEIVERS

Certain precautions should be taken when servicing solid-state (transistorized) CB equipment:

1. Observe battery polarity. Unless the unit has a protector circuit, reverse polarity can destroy transistors.

2. Always connect a 50-ohm dummy load to the transceiver antenna terminal. Turning the transmitter on with no antenna load can result in transistor destruction.

3. Do not use an ohmmeter for testing transistor circuits unless you are sure the meter battery (within the ohmmeter) polarity is not reversed in regard to transistors.

4. **Do not unsolder** transistors for testing. Test them in the circuit.

5. When soldering transistor leads or components connected to transistors, avoid overheating which can destroy a transistor. Use a low-wattage soldering iron and pliers to dissipate the heat.

TRANSMITTER OSCILLATOR (Transistor)

The oscillator (Fig. 3-6) generates an unmodulated RF signal, usually at the channel frequency. The signal frequency is determined by the crystal (Y). Generally, a switch is included (not shown) which selects the appropriate crystal for the selected channel. The output signal is developed across the primary of RF transformer T and fed from its secondary to the next stage.

Checkout Steps

1. Measure the RF output voltage (with a VTVM and an RF probe) at the secondary of T. Adjust T for maximum meter indication. Then, adjust T for a slightly lower output on the gentle slope of the resonance curve.

2. Key the transmitter on and off. The RF output should be indicated every time.

MODULATED TRANSMITTER BUFFER AMPLIFIER

The oscillator signal is fed through T1 to the base of transistor Q in Fig. 3-7. It's amplified output is developed across L2 and fed to the next stage through C4.

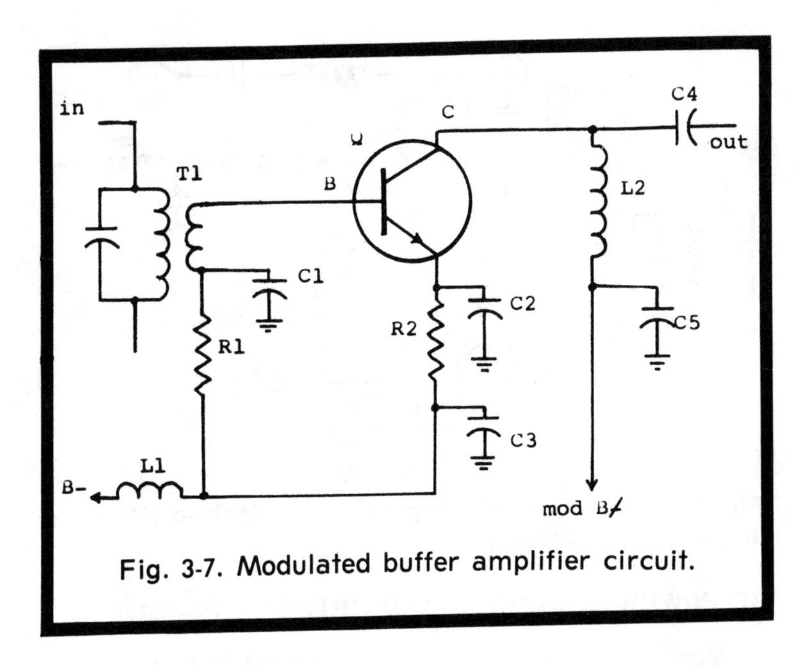

Fig. 3-7. Modulated buffer amplifier circuit.

Checkout Steps

1. Measure the DC voltage across R2. Short B-E of the transistor. The voltage should drop.

2. Measure the RF voltage across L2 with a VTVM equipped with an RF probe. The voltage should vary when talking into the mike.

Defect	Symptom	Evidence
Transistor open	Won't transmit	No DC voltage across R2
Transistor shorted	Won't transmit	No DC voltage across C-E of Q
L2 open	Won't transmit	No DC voltage across R2
L1 open	Won't transmit	No DC voltage across C3
C2 open	Low output	New capacitor restores operation

Chart 3-7. Modulated buffer troubleshooting hints.

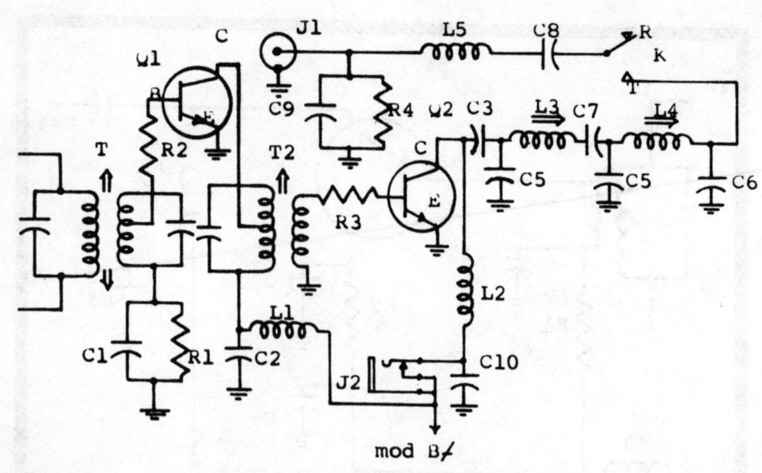

Fig. 3-8. Schematic of a typical transistorized RF power amplifier and driver.

RF POWER AMPLIFIER AND DRIVER (Transistor)

Signal from the oscillator (Fig. 3-8) is fed through T1 to the base of Q1, whose output is fed through T2 to the base of Q2. The output of Q2 is fed through C3 (a doubler pi network), T-R relay contacts, C8 and L5 to antenna connector J. The Q2 emitter current can be measured by plugging a DC ammeter or VOM into J2.

Defect	Symptom	Evidence
Q1 shorted	Won't transmit	No DC voltage across C-E of Q1
Q2 shorted	Won't transmit	No DC voltage across C-E of Q2
L1 open	Won't transmit	No DC voltage across C2
L2 open	Won't transmit	No DC voltage across C-E of Q2
C3, C7, C8 or L5 open	Won't transmit	No RF voltage at J

Chart 3-8. Troubleshooting data, transistorized RF power amplifier and driver.

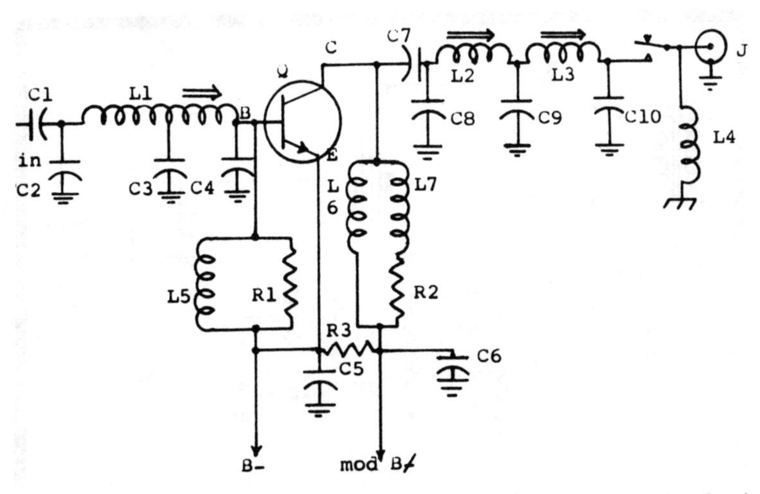

Fig. 3-9. RF power amplifier using input and output pi networks.

Checkout Steps

1. Connect a No. 47 lamp or RF wattmeter to J. Turn the transmitter on. Power output should be indicated.

2. Same as above. Talk into the mike. The power output should rise.

Defect	Symptom	Evidence
Transistor shorted	Won't transmit	No DC voltage across C-E of Q
C2, C3 or C4 shorted	Won't transmit	No DC voltage across C4
C7 shorted	Won't transmit	DC voltage present across C8
C6 shorted	Won't transmit	No DC voltage across C6
C8, C9 or C10 open or shorted	Low output	L2 and-or L3 won't tune

Chart 3-9. Troubleshooting tips for an RF power amplifier with pi networks at the input and output.

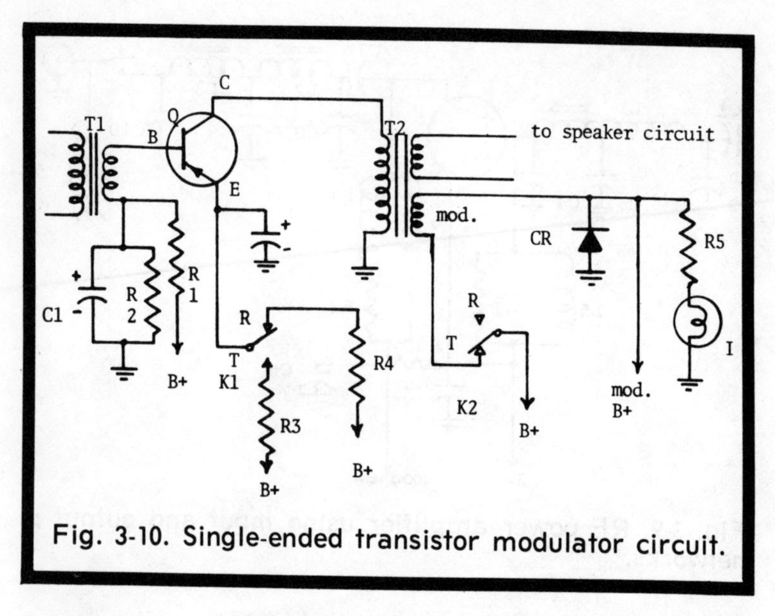

Fig. 3-10. Single-ended transistor modulator circuit.

3. Connect a VOM (set to read milliamperes) to J2 through a phone plug (minus lead to tip). Turn the transmitter on. The meter should not indicate more than 362 milliamperes with 13.8 volts input.

RF POWER AMPLIFIER WITH PI-NETWORK INPUT AND OUTPUT

Signal from the buffer amplifier is fed through C1 and L1 to the base of transistor Q (Fig. 3-9). The amplified output is fed through C7, the double pi network (C8-C9-C10-L3), the T-R relay K contacts to antenna connector J. Modulated B+ varies the power output with modulation.

Checkout Steps

1. Connect a No. 47 lamp or RF wattmeter to J. Turn the transmitter on. Power output should be indicated.

2. Same as above. Also talk into the mike. The power output should rise.

SINGLE-ENDED TRANSISTOR MODULATOR

Transistor Q in Fig. 3-10 functions as a common-emitter Class B power amplifier. The base is biased through voltage

Defect	Symptom	Evidence
Transistor or primary of T2 open	Won't modulate-receive	No DC voltage across C and common ground
Transistor shorted	Won't modulate-receive	No DC voltage across C-E
C2 shorted	Won't modulate-receive	No DC voltage across C2
R3 open	Won't modulate	No DC voltage across C2 when set to transmit
R4 open	Won't receive	No DC voltage across C2 when set to receive
CR shorted	Lamp won't light	No DC voltage across CR when set to transmit

Chart 3-10. Single-ended transistor modulator troubleshooting data.

divider R1-R2. The AF signal is fed from the base through T1 and from the collector to the speaker and modulated stages through T2. Diode CR prevents overmodulation. Lamp I lights when transmitting and brightens with modulation. **Note:** Although R1 goes to B+, the base is nevertheless negative with respect to the emitter.

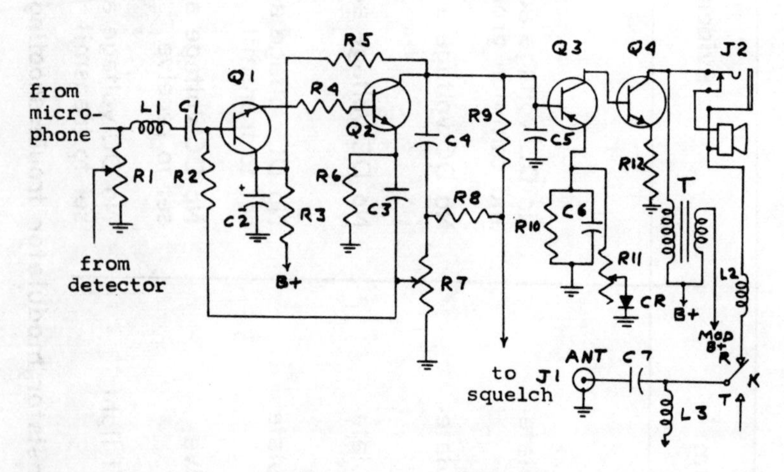

Fig. 3-11. Solid-state direct-coupled modulator-AF amplifier

Defect	Symptom	Evidence
Q4 shorted	Won't operate	High DC voltage across R12
Q4 open	Won't operate	No DC voltage across R12
L2 or speaker open	Won't receive	No sound heard in earphones plugged into J2
T secondary open	Won't modulate	No DC voltage from X to ground

Chart 3-11. Troubleshooting data, solid-state direct-coupled modulator-AF amplifier.

Checkout Steps

1. Apply an AF test signal to the primary of T1. Measure the AF voltage across the modulator winding of T2 with the unit set to receive. Then set the unit to transmit. The AF voltage should rise.

2. Measure the DC voltage across C2. Momentarily short B-E. The voltage should drop.

3. Measure the DC voltage across CR. It should be zero in receive and more than 12 volts in the transmit condition.

SOLID-STATE DIRECT-COUPLED MODULATOR-AF AMPLIFIER

All four transistors are direct coupled (Fig. 3-11). Therefore, resistor values are critical. All transistors are NPN types, except Q3 which is a PNP type. Both the mike output and receiver output are amplified by all four stages. The 25-ohm speaker is directly in series with the collector of Q4. Transformer T provides the required AF modulating voltage.

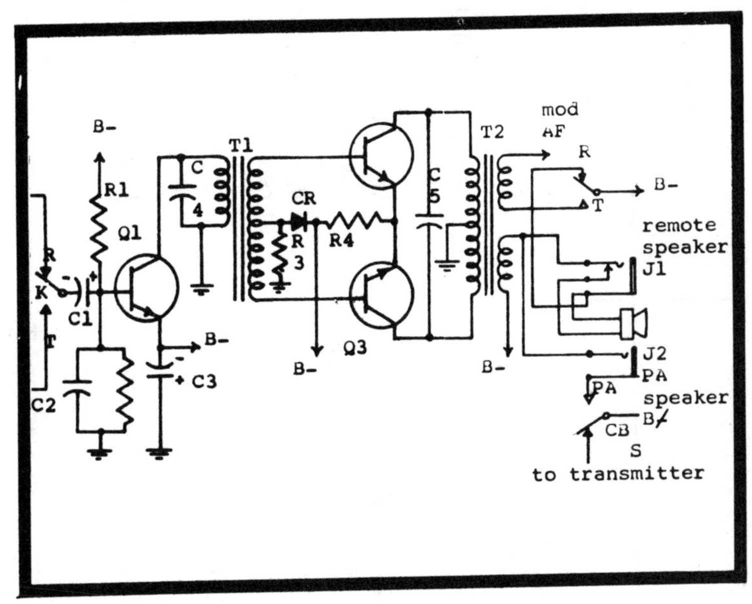

Fig. 3-12. Schematic of a solid-state modulator with push-pull output.

Defect	Symptom	Evidence
Q1 open	Won't modulate-receive	No DC voltage across C4
Q1 shorted	Won't modulate-receive	High DC voltage across C4
Q2 or Q3 shorted	Won't modulate-receive	High DC voltage across R4
C3 shorted	Won't modulate-receive	No DC voltage across C3
Diode CR open	Distorted modulation	High DC voltage across CR
Diode CR shorted	Low modulation	High DC voltage across R3

Chart 3-12. Troubleshooting hints for a solid-state push- pull modulator.

Checkout Steps

1. Apply an AF test signal to L1. Connect the vertical input of the oscilloscope to C of Q4 and common ground. Observe the waveform. If not clean, adjust R7 and R11 carefully. **Note:** Refer to the applicable service manual for DC voltage checks.

SOLID-STATE MODULATOR WITH PUSH-PULL OUTPUT

AF signals fed through C1 are amplified by transistor Q1 which drives Q2 and Q3 in push-pull through transformer T1 (Fig. 3-12). The AF voltage across the top secondary of T2 alternately bucks and boosts the B- voltage fed to the modulated stages. When switch S is set to PA, the transmitter is inoperative.

Checkout Steps

1. Apply an AF test signal to C1. Connect an external speaker to J2. Set S to CB; the test tone should not be heard when the unit is set to transmit. Then set S to PA; the test tone should now be heard when the unit is set to transmit.

2. Measure the DC voltage across C5. It should be zero or almost. If not, Q2 and Q3 are not balanced.

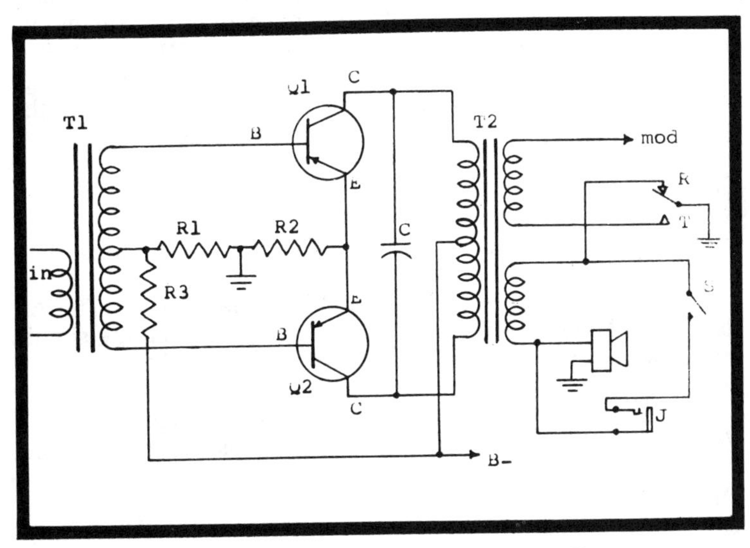

Fig. 3-13. Class B push-pull modulator-AF power amplifier using PNP transistors.

Defect	Symptom	Evidence
Q1 or Q2 shorted	Won't operate	DC voltage present across C
Q1 or Q2 open	Distortion	Full DC supply voltage across C-E of Q1 and Q2
R2 open	Won't operate	No DC voltage across C-E of Q1 and Q2
R1 open	Distortion	Excessive DC voltage across R1
R3 open	Low output or won't operate	No DC voltage across R1

Chart 3-13. Class B push-pull modulator troubleshooting data.

CLASS B PUSH-PULL MODULATOR-AF POWER AMPLIFIER EMPLOYING PNP TRANSISTORS

Transistors Q1 and Q2 function as a Class B push-pull amplifier (Fig. 3-13). When the base of Q1 is fed a negative-going signal, its collector current rises. At the same time a positive-going signal is fed to the base of Q2, causing its collector current to fall. When the signal to Q1 is positive-going, just the opposite happens. The resulting AF voltage developed across the primary of T2 is stepped down to the speaker and stepped up to the modulated stages.

Checkout Steps

1. Measure the DC voltage across C. It should be zero.

2. Apply an AF test signal to the primary of T1. The AF tone should be heard in the speaker and an AF voltage should be present across the modulator secondary of T2.

3. Same as above, but measure the DC voltage across R2. Voltage should increase as the AF test signal level is raised.

PREAMP CERAMIC MIKE

Transistor Q in Fig. 3-14 functions as a common-emitter AF amplifier and is cut into the circuit by the contacts of T-R relay K. The output signal is developed across R7. The emitter is at above ground potential for DC. C2, L and R2 provide frequency compensation.

Checkout Steps

1. Apply an AF test signal to R1. Connect headphones (through a capacitor) across R7 and listen for the AF tone.

2. Measure the DC voltage across R7. Momentarily short B-E; the voltage should drop.

RECEIVER PENTODE RF AMPLIFIER WITH PI-NETWORK INPUT (SHARED WITH TRANSMITTER)

RF signals from antenna connector J are fed through a pi network (C2-C3-L3) and C4 to the grid of the tube. The amplified output is fed through RF transformer T1 to the mixer. Gain is automatically controlled by the AVC voltage fed through R1.

Checkout Steps

1. Apply an RF test signal to J. Measure the DC voltage across C8. The voltage should rise as the test signal level increases.

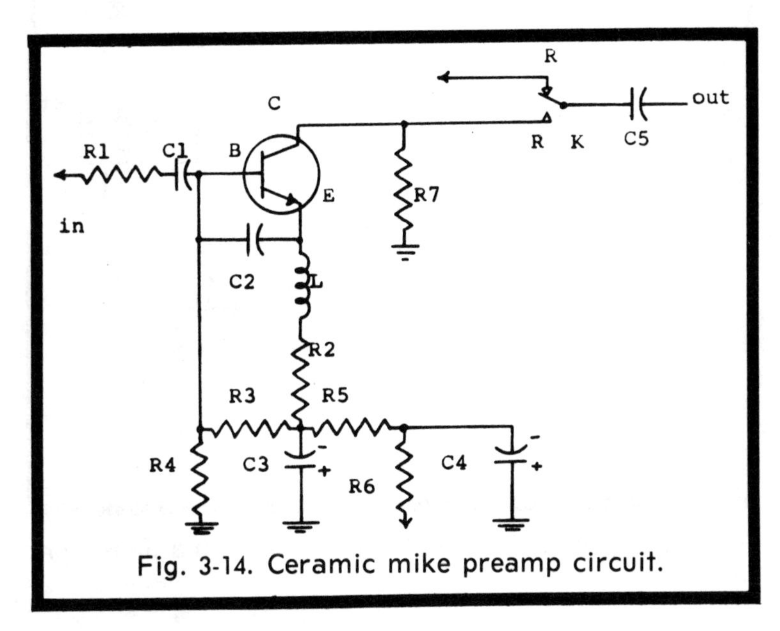

Fig. 3-14. Ceramic mike preamp circuit.

2. Same as above, except short C5. The voltage should drop.

3. Same as above, except short R2. The voltage should drop.

Defect	Symptom	Evidence
Transistor open	Won't modulate	No DC voltage across R7
Transistor shorted	Won't modulate	No DC voltage across C-E
C1 or R1 open	Won't modulate	No AF voltage across R7
C3 or C4 shorted	Won't modulate	No DC voltage across R4

Chart 3-14. Mike preamplifier troubleshooting data.

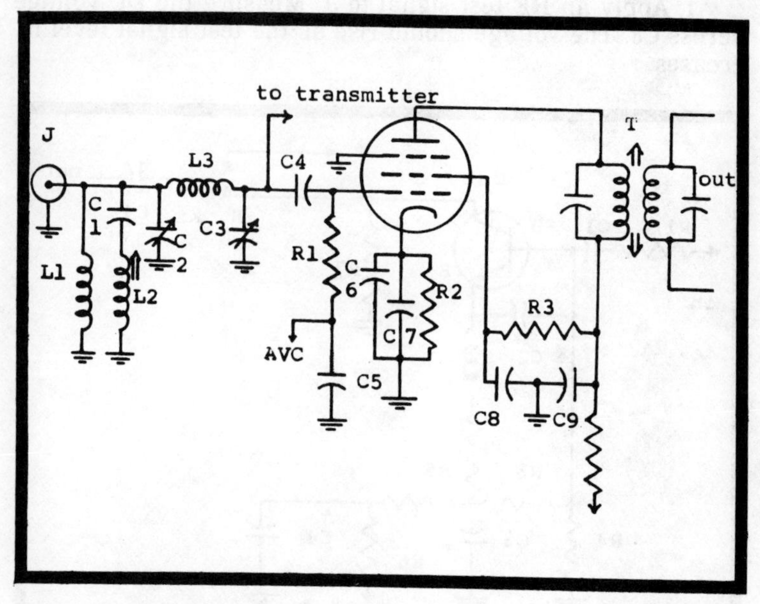

Fig. 3-15. Pentode RF amplifier circuit with a pi-network input.

Defect	Symptom	Evidence
C4 open	Noise, no signal	New capacitor restores operation
L3 open	Noise, no signal	Shorting L3 restores operation
C8 shorted	Won't operate	No DC voltage across C8
C9 shorted	Won't operate	No DC voltage across C9
R3 open	Won't operate	No DC voltage across C8
R4 open	Won't operate	No DC voltage across C9

Chart 3-15. Troubleshooting tips, RF amplifier with a pi-network input.

Chapter 4

Troubleshooting Receiver Stages

The same logical transmitter troubleshooting approach is valid for receiver servicing, too. Since the audio section is usually common to both transmitter and receiver, it can be assumed operational in the receiver if the transmitter is known to be modulating properly. Therefore, by checking from the detector stage back through the IF stages and then the mixer, oscillator and RF amplifier, a faulty stage may be found rapidly.

The use of the signal injector is a great time saver and many troubles can be found without the need of a signal generator and VTVM. However, when using a signal generator, always keep its RF output as low as possible to avoid false indications that come when a circuit is overloaded by too strong a signal. Also care should be taken not to load down a stage by directly coupling a generator or multimeter to it. Whenever possible, use a blocking capacitor on the test probe.

CASCODE RF AMPLIFIER

The two triode tubes in Fig. 4-1 are connected in cascode configuration. They are in series for DC and direct coupled. Signals from the antenna are fed to the grid of V1, whose output is fed to the cathode of V2 which functions as a grounded-grid amplifier. The output of V2 is fed through RF transformer T to the mixer.

Checkout Steps

1. Measure the DC voltage from the cathode to the plate of V2 and adjust R3. Voltage should vary.

2. Apply a strong RF test signal to T1. Measure the DC voltage across C3. It should decrease as the test signal level increases.

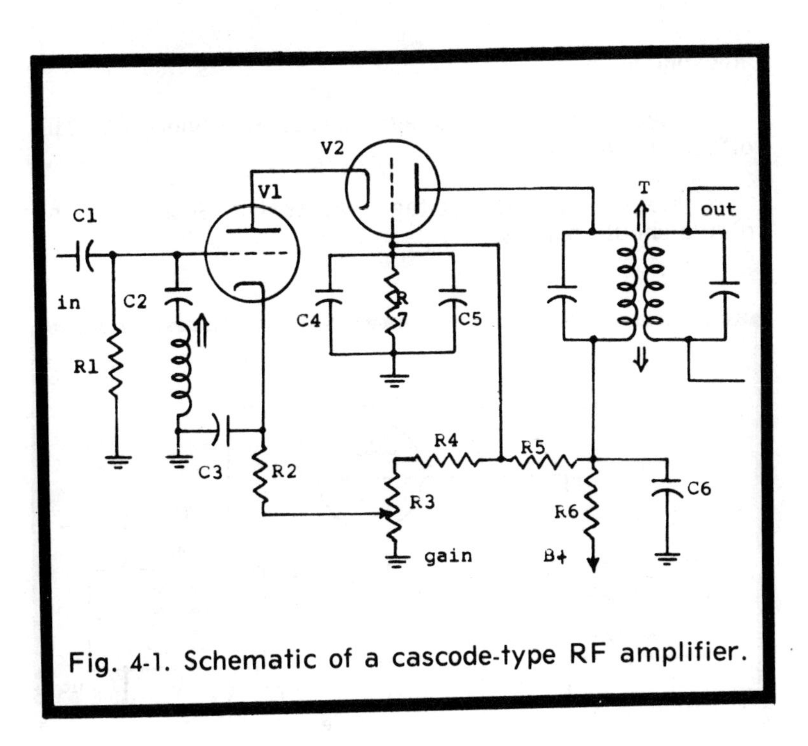

Fig. 4-1. Schematic of a cascode-type RF amplifier.

Defect	Symptom	Evidence
C6 shorted	Won't receive	No DC voltage across C6
R6 open	Won't receive	No DC voltage across C6
C4 or C5 shorted	Won't receive	No DC voltage across R7
R2 open	Won't receive	No DC voltage across C3

Chart 4-1. Troubleshooting tips, cascode RF amplifier.

TRIODE FIRST MIXER STAGE

The on-channel signal from the RF amplifier is fed to the grid of the mixer tube through RF transformer T1 (Fig. 4-2). A signal from the local oscillator is also fed to the grid through C1. Their sum (or difference) frequency signal is fed through IF transformer T2 to the second mixer.

Checkout Steps

1. Measure the DC voltage across R3. Short C2. The voltage should rise.

2. Measure the DC voltage across R2. Short R1. The voltage should not change.

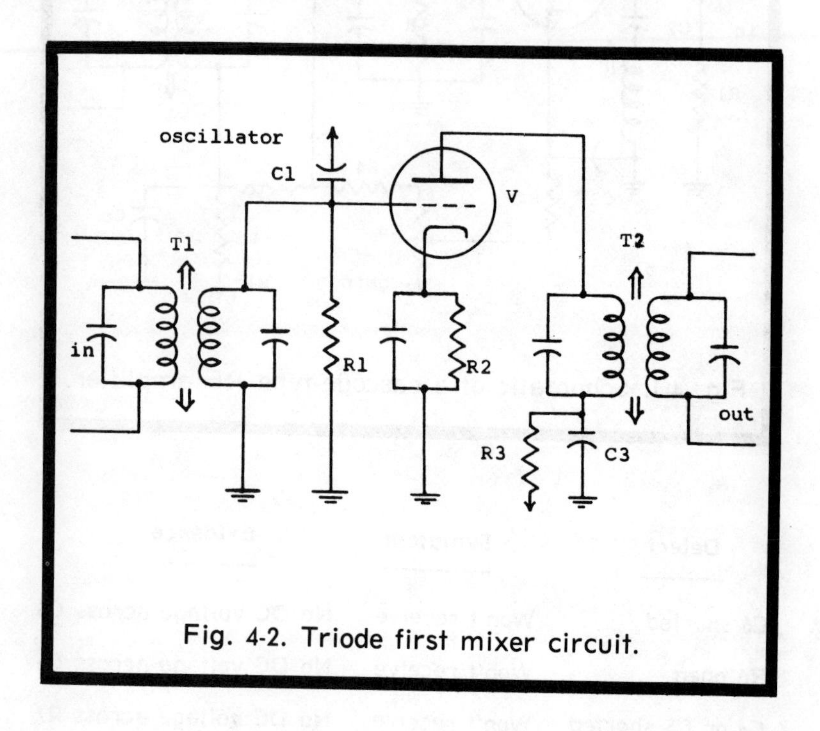

Fig. 4-2. Triode first mixer circuit.

Defect	Symptom	Evidence
C1 open	Noisy reception	New capacitor restores operation
R2 open	No operation	No DC voltage across R3
C3 shorted	No operation	No DC voltage across C3

Chart 4-2. Typical first mixer trouble.

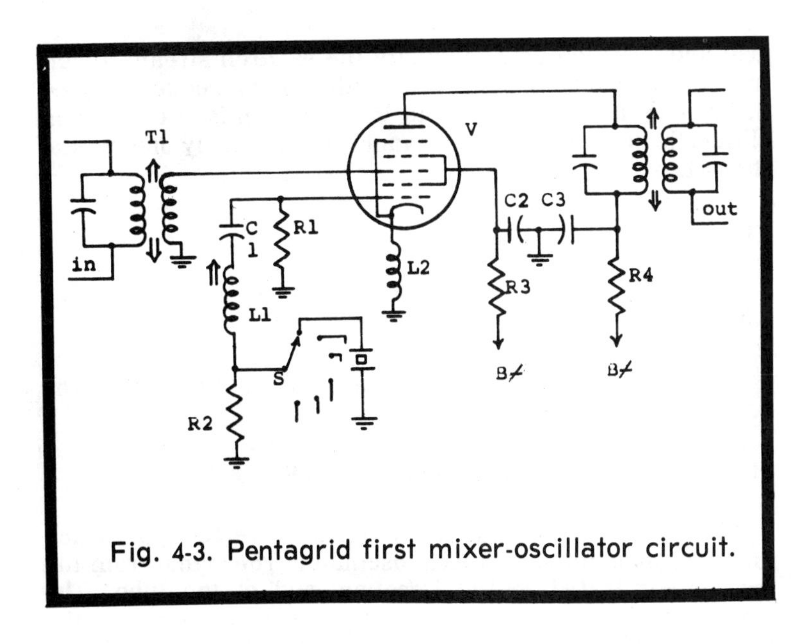

Fig. 4-3. Pentagrid first mixer-oscillator circuit.

Defect	Symptom	Evidence
C3 shorted	Won't receive	No DC voltage across C3
C4 shorted	Won't receive	No DC voltage across C4
C1 shorted	Won't receive	No DC voltage across R1
C1 open	Won't receive	New capacitor restores operation
L2 open	Won't receive	No DC voltage across R4

Chart 4-3. Pentagrid first mixer-oscillator troubleshooting hints.

PENTAGRID FIRST MIXER-OSCILLATOR

On-channel signals (Fig. 4-3) are fed through RF transformer T1 to the injection grid of the pentagrid converter tube. The control grid, cathode and screen grid (grounded for RF through C2) are used in an oscillator circuit. The control grid and injection grid modulate the electron stream at different frequencies. Their sum (or difference) frequency is fed through IF transformer T2 to the second mixer or IF amplifier. Switch S selects the channel crystals (only one crystal shown).

Checkout Steps

1. Measure the DC voltage across R4. Short R1. The voltage should rise.

2. Measure the DC voltage across R4. Short R3. The voltage should rise.

SECOND MIXER PENTAGRID CONVERTER

In Fig. 4-4 the tube is a pentagrid converter type which functions as a mixer and local oscillator. The signal from the first mixer is fed to the injection grid of the tube. The frequency-determining crystal is connected between the control grid and the screen grid to form a Pierce oscillator. The grid modulates the electron stream at the oscillator frequency. The injection grid also modulates the electron stream, but at the frequency of the incoming signal. Both frequencies plus their sum and difference frequencies are present across R4 and are fed through C4 to a selectivity filter (not shown) tuned to the sum or difference frequency.

Checkout Steps

1. Apply a strong RF test signal to the primary of T. Measure the DC voltage across R1. The voltage should rise as the test signal level increases.

2. Same as above, except measure the DC voltage across R4. Then short R1. The voltage should rise.

3. Measure the DC voltage across R2. It should be present with the crystal in place.

TWO-STAGE IF AMPLIFIER

The IF signal is fed through selectivity filter F to the control grid of V1, from V1 through IF transformer T1 to V2, and from V2 through T2 to the detector (Fig. 4-5).

Checkout Steps

1. Apply an RF test signal to C1. Measure the DC voltage across C6. The voltage should rise as the test signal level is increased.

2. Measure the DC voltage across C2. Short R2. The voltage should drop.

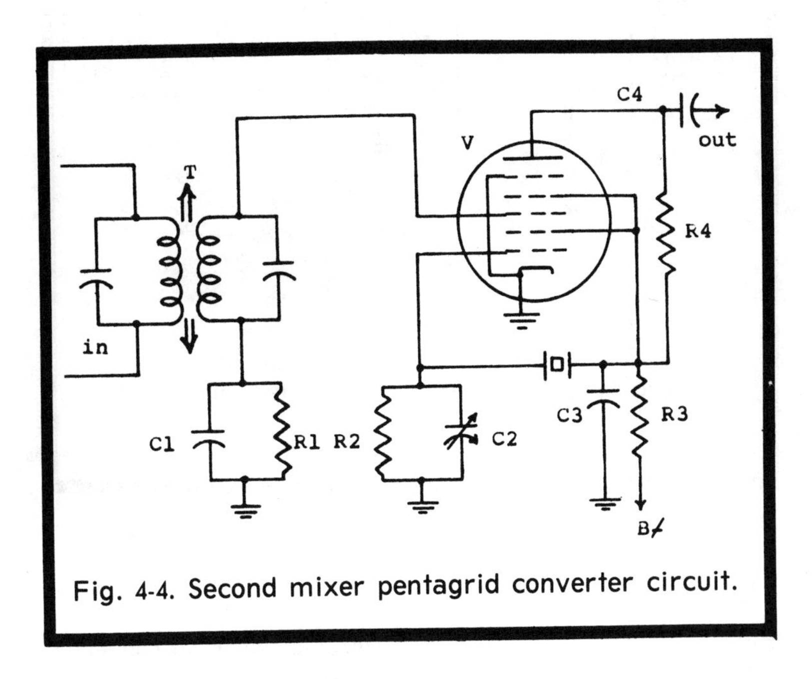

Fig. 4-4. Second mixer pentagrid converter circuit.

Defect	Symptom	Evidence
C3 shorted	Won't receive	No DC voltage across C3
R3 open	Won't receive	No DC voltage across C3
R4 open	Won't receive	No DC voltage from plate to ground
Crystal defective	Won't receive	No DC voltage across R2
R2 value changed	Won't receive	No DC voltage across R2

Chart 4-4. Second mixer pentagrid troubleshooting tips.

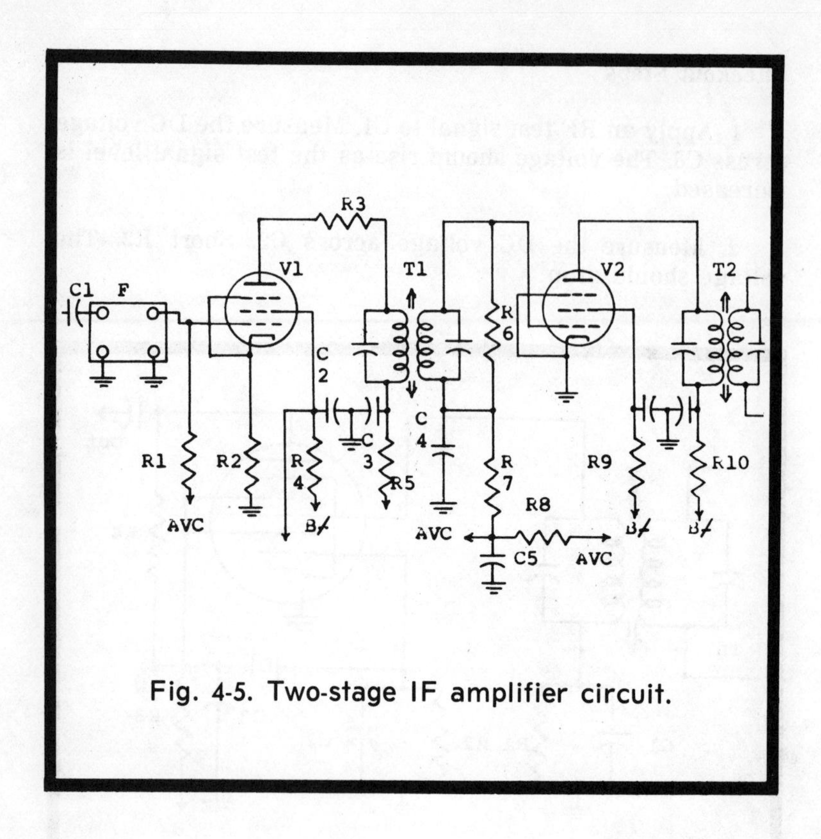

Fig. 4-5. Two-stage IF amplifier circuit.

Defect	Symptom	Evidence
Filter open or shorted	Won't receive	Receives when C1 connected to R1
R2 open	Won't receive	No DC voltage across R5
C2 shorted	Won't receive	No DC voltage across C2
C3 shorted	Won't receive	No DC voltage across C3
C4 open	Low sensitivity	Corrected by new capacitor
C6 shorted	Won't receive	No DC voltage across C6
C7 shorted	Won't receive	No DC voltage across C7

Chart 4-5. Troubleshooting tips, two-stage IF amplifier.

3. Measure the DC voltage across C6. Short C4. The voltage should drop.

DIODE TUBE DETECTOR & SERIES NOISE LIMITER

Diode V1 is the detector in Fig. 4-6. Its output is an AF signal and a negative DC voltage which is developed across R3

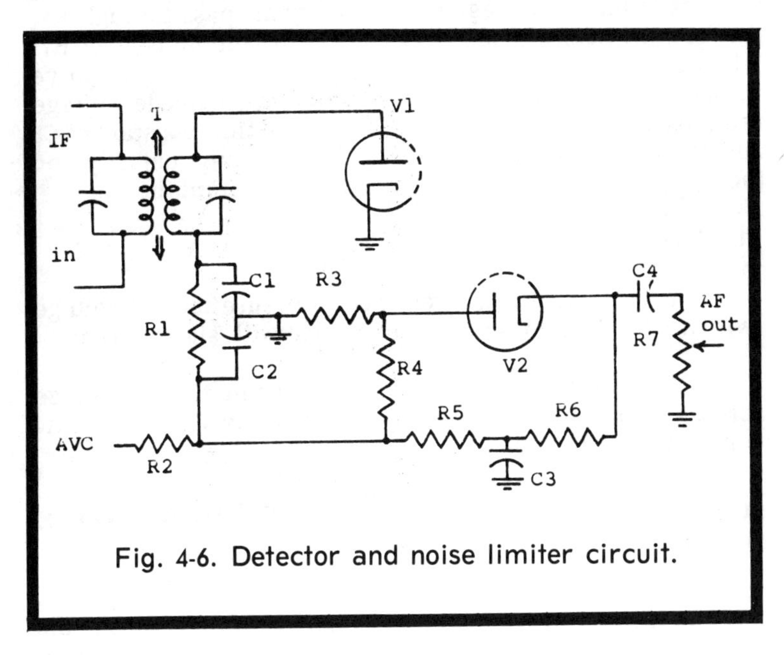

Fig. 4-6. Detector and noise limiter circuit.

Defect	Symptom	Evidence
R1 open, C1 or C2 shorted	Won't operate	No DC voltage across C2 when signal present
R4 open	Won't operate	No DC voltage across R3
R3, R5 or R6 open	No noise limiting	No DC voltage across V2 plate and cathode
C5 shorted	No noise limiting	No DC voltage across C5
C5 open	No noise limiting	Try new capacitor

Chart 4-6. Detector and noise limiter troubles.

and R4. C1 and C2 bypass RF. When a signal is received, the cathode of diode V2 becomes more negative than its plate (making the plate positive with respect to the cathode) because all of the DC voltage across R3 and R4 is fed to the cathode through R5 and R6, and the plate is connected to a lower negative potential point, the junction of R3 and R4. C3 charges to the DC voltage level. AF signals pass through R4, the forward-biased diode (V2) and C4 to volume control R7. When a noise pulse is received the instantaneous negative voltage at the plate of V2 rises. But the cathode voltage remains at its previous level because of the time constant of C3 and R5. Hence, diode V2 is reverse-biased momentarily and the noise pulse doesn't get through to the AF amplifier.

Checkout Steps

1. With a signal being received, measure the DC voltage (with a VTVM) across C3. It should rise with signal strength.

2. Momentarily short R4 while measuring the DC voltage across C3. The voltage should not change if the short across R4 is not present too long.

SCREEN VOLTAGE CONTROLLED DIODE GATE SQUELCH

The squelch circuit in Fig. 4-7 responds to variations in the screen voltage of an AVC-controlled IF amplifier stage (V1). When no signal is being received, AVC voltage is low and screen current is high. When a signal is received, AVC voltage rises and reduces screen current. Therefore, the screen voltage (across C2) rises. With no signal received, diode V2 is reverse-biased and blocks the passage of AF signals from C4 to C5. Its cathode is more positive than its plate. When the screen voltage rises because of an incoming signal, the voltage across squelch control R6 rises and makes the plate of V2 more positive than its cathode. It now conducts and passes AF signals.

Checkout Steps

1. Apply an RF test signal to the grid of V1. Measure the DC voltage across R6 set in the fully unsquelched position. The DC voltage should rise as the test signal level increases.

2. Measure the DC voltage across R10 (with a VTVM) and adjust R6. When the voltage is present, the squelch is open.

Note: Changes in resistor values can affect operation.

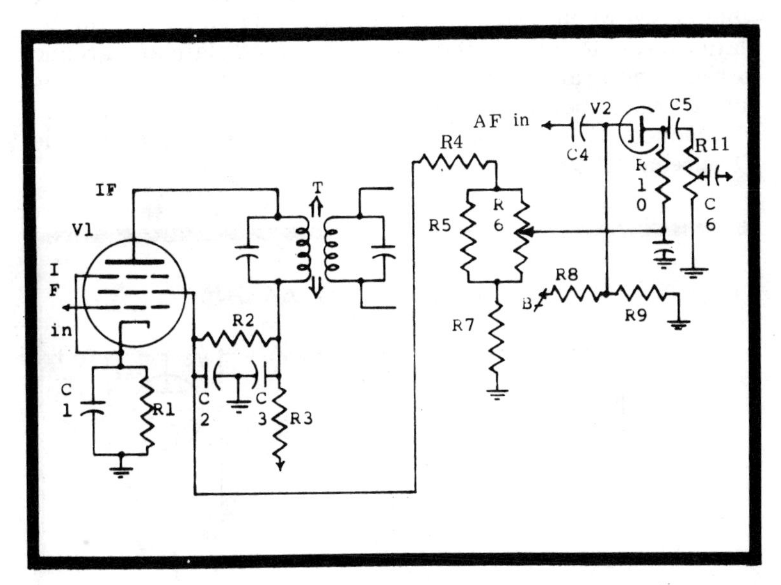

Fig. 4-7. Screen voltage controlled diode gate squelch circuit.

Defect	Symptom	Evidence
R4 open or C7 shorted	Squelch won't open	No DC voltage across R6
R7 open	Squelch won't quiet	DC voltage across C7 can't be varied with R6
R8 open	Squelch won't quiet	No DC voltage across R9
R9 open	Squelch won't open	Full DC supply voltage across R9
C5 leaky or shorted	Squelch won't open	DC voltage present across R11

Chart 4-7. Diode gate squelch trouble tips.

PENTODE DC AMPLIFIER SQUELCH

AF amplifier V2 (Fig. 4-8) is normally biased to cutoff by the application of a positive voltage to its cathode through R4 and R8. Its grid is negative with respect to the cathode. Squelch control tube V2 senses a negative DC AGC voltage which rises when a signal is received. A signal causes the V2 plate current to fall and the voltage between plate and ground to rise. The grid of V2 senses V1's plate voltage. When this voltage rises (becomes more positive) it offsets the bias and V2 conducts. R2 is the squelch control which is used to vary the V1 screen voltage.

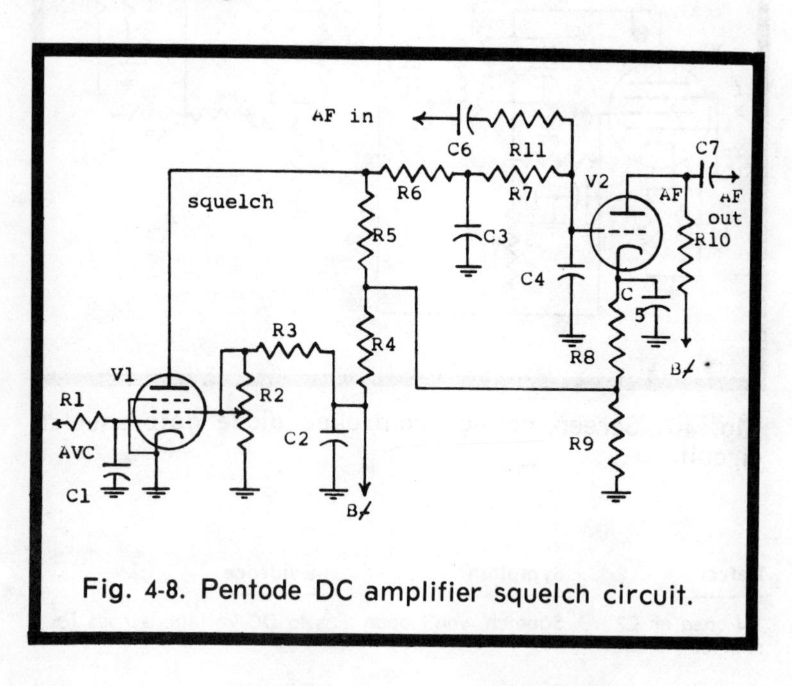

Fig. 4-8. Pentode DC amplifier squelch circuit.

Defect	Symptom	Evidence
R3 open	Squelch won't open	No DC voltage across R3
R4 open	Squelch won't quiet	No DC voltage across C3
C3 or C4 shorted	Squelch won't open	No DC voltage across C4

Chart 4-8. Troubleshooting tips for a pentode DC amplifier squelch circuit.

Checkout Steps

1. Measure the DC voltage across R8. Adjust R2. The voltage should be present when the receiver is unsquelched, none when squelched.

2. Apply an RF test signal to the receiver input and check the signal level required to open the squelch at the threshold.

GROUNDED-BASE TRANSISTOR RF AMPLIFIER WITH AGC

Signals from the antenna connector (J) are fed through L, C2, the normally closed contacts of relay K to the primary of

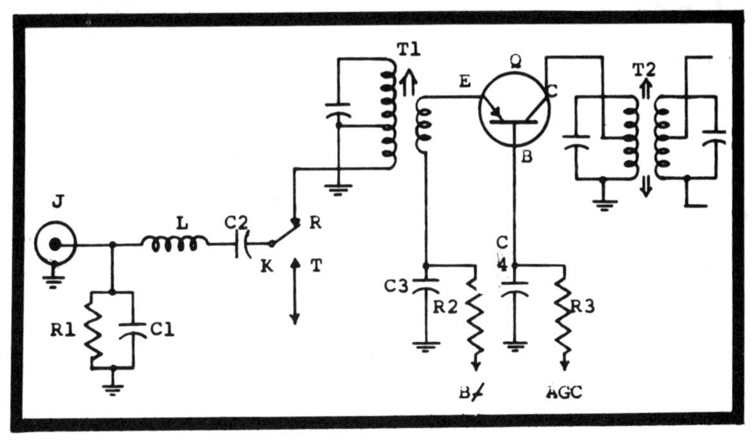

Fig. 4-9. Grounded-base RF amplifier circuit with AGC.

Defect	Symptom	Evidence
C3 shorted	Won't receive	No DC voltage across C3
R2 open	Won't receive	No DC voltage across C3
C4 open	Noisy reception	New capacitor restores operation
Transistor shorted	Won't receive	No DC voltage across C3
Open transistor	Won't receive	No DC voltage across R2

Chart 4-9. Trouble hints for the grounded-base RF amplifier.

RF transformer T1 (Fig. 4-9). The collector of transistor Q is fed from the secondary of T1. The output of Q is fed through RF transformer T2 to the mixer.

Checkout Steps

1. Measure the DC voltage across R2. Short B-E of the transistor. The voltage should drop.

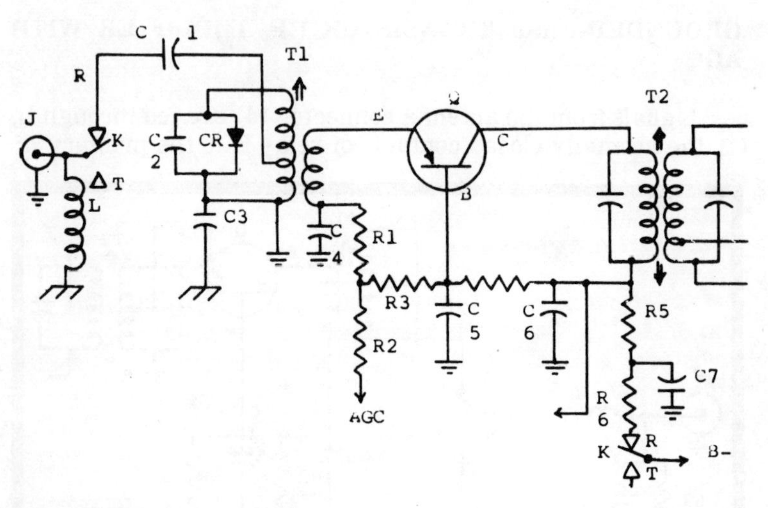

Fig. 4-10. RF amplifier circuit with AGC and overload protection.

Defect	Symptom	Evidence
Transistor shorted	Won't receive	No DC voltage across C-E of Q
C7 shorted	Won't receive	No DC voltage across C7
C5 open	Insensitive	New capacitor restores operation
C1 open *	Noise	New capacitor restores operation

* Poor contact of antenna relay contacts can cause the same problem.

Chart 4-10. Troubleshooting tips for the RF amplifier with AGC and overload protection.

2. Apply an RF test signal to J. Measure the voltage across R2. It should drop as the test signal level increases.

RF AMPLIFIER WITH AGC AND OVERLOAD PROTECTION

Signals are fed from antenna connector J through the normally closed contacts of T-R relay K and C1 to a low-impedance tap on the primary of RF transformer T1 (Fig. 4-10). The secondary of T1 feeds the emitter of transistor Q, whose output is fed through RF transformer T2 to the mixer. Diode CR protects the transistor from static and strong signals.

Checkout Steps

1. Measure the DC voltage across C-E of the transistor. Short B-E. The voltage should rise.

2. Disconnect one diode lead. Measure the resistance of the diode. Then reverse the ohmmeter leads. One way, the resistance should be extremely high, and very low or zero the other way.

FET RF AMPLIFIER

Signals are fed from antenna relay K to T1 (Fig. 4-11). The secondary tap on T1 feeds the FET whose output is fed through transformer T2 to the mixer. Diode CR1 protects the FET from overloading on strong signals.

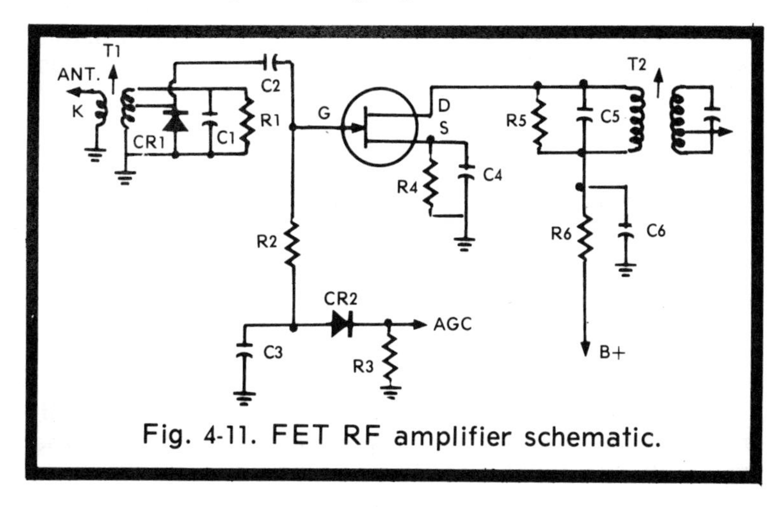

Fig. 4-11. FET RF amplifier schematic.

Defect	Symptom	Evidence
FET shorted	Won't receive	No DC voltage across D-S of Q
C6 shorted	Won't receive	No DC voltage across C6
C4 open	Insensitive	New C4 restores signal
C2 open	No signal but noisy	New C2 restores operation
CR1 shorted	No signal	New CR1 restores operation

Chart 4-11. Trouble tips, FET RF amplifier.

Checkout Steps

1. Measure the DC voltage across D-S of the FET. Short G-S and the voltage will rise.

2. Disconnect one side of CR1. Measure the resistance of CR1. Reverse the meter leads and again read the resistance. One reading should be very high and the other reading very low.

FET MIXER

The on-channel signal from the RF amplifier is fed through T1 to the gate of Q and mixed with the local oscillator signal. The resulting IF signal is fed through T2 to the second mixer (Fig. 4-12).

Checkout Steps

1. Check the voltages on the FET.

2. Check the incoming RF signals and local oscillator signals with an RF voltmeter.

MIXER DIODE

The on-channel signal from an RF amplifier is fed to the anode of diode CR through RF transformer T1 (Fig. 4-13). The local oscillator signal is fed from T2 through L1 to the diode cathode. The resulting IF signal is fed through C1 and L2 to the second mixer.

Checkout Steps

1. With the receiver turned off, disconnect one end of the diode and measure its resistance. Then reverse the ohmmeter leads. It should show very high resistance one way and short circuit or very low resistance the other way.

MULTI-CHANNEL FIRST MIXER-OSCILLATOR

On-channel signals are fed from the RF amplifier through RF transformer T1 to the base of transistor Q, which functions as a crystal-controlled autodyne converter (Fig. 4-14). S selects the channels (only one crystal shown). The local oscillator circuit utilizes T2. The resulting IF signal is fed through IF transformer T3 to the IF amplifier.

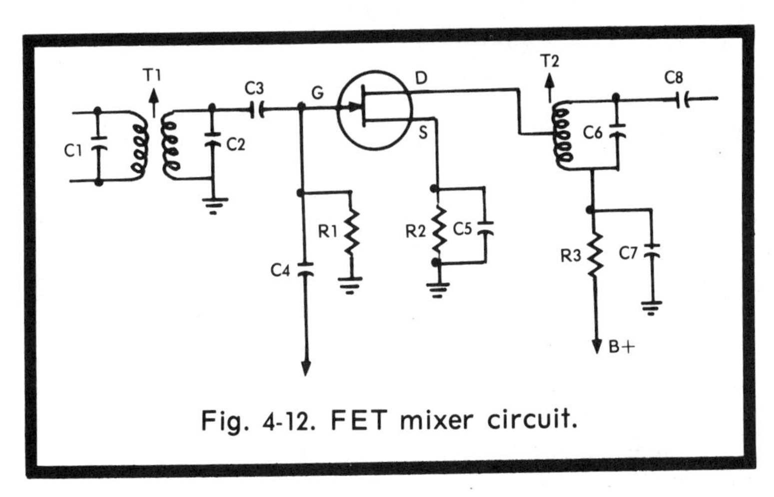

Fig. 4-12. FET mixer circuit.

Defect	Symptom	Evidence
FET shorted	Won't receive	New FET restores signal
C4 open	Won't receive	New C4 restores local oscillator signal
C7 shorted	Won't receive	No DC voltage at D of FET

Chart 4-12. FET mixer trouble hints.

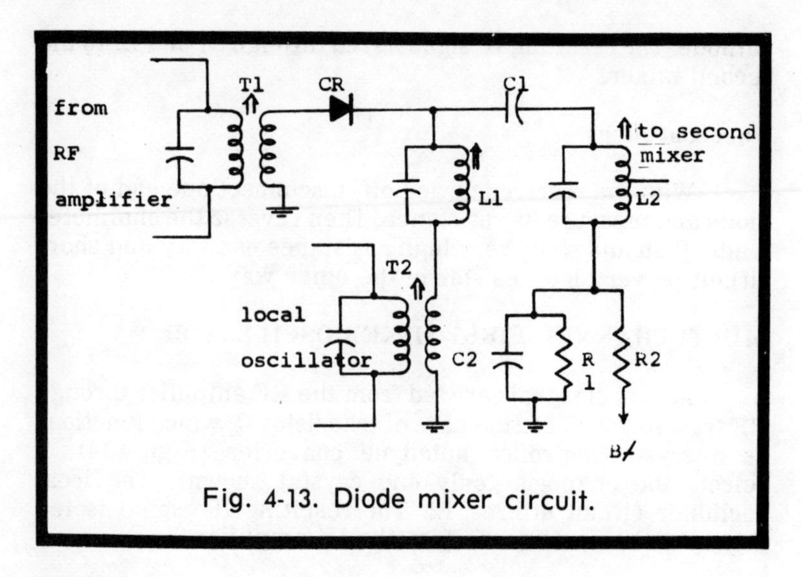

Fig. 4-13. Diode mixer circuit.

Defect	Symptom	Evidence
Diode open	Won't receive	New diode restores operation
C1 open	Noisy reception	New capacitor restores operation

Chart 4-13. Diode mixer trouble hints.

Checkout Steps

1. Measure the DC voltage across R4. Short B-E of Q. The voltage should drop.

2. Measure the DC voltage across R4. Short C1. The voltage should drop.

SECOND MIXER

Transistor Q in Fig. 4-15 functions as both a mixer and local oscillator using an LC resonant circuit (T2) instead of a crystal. The signal from the first mixer is fed to the base of Q through T1. The collector is coupled to the emitter through T2. The resulting IF signal is fed to the IF amplifier through T3.

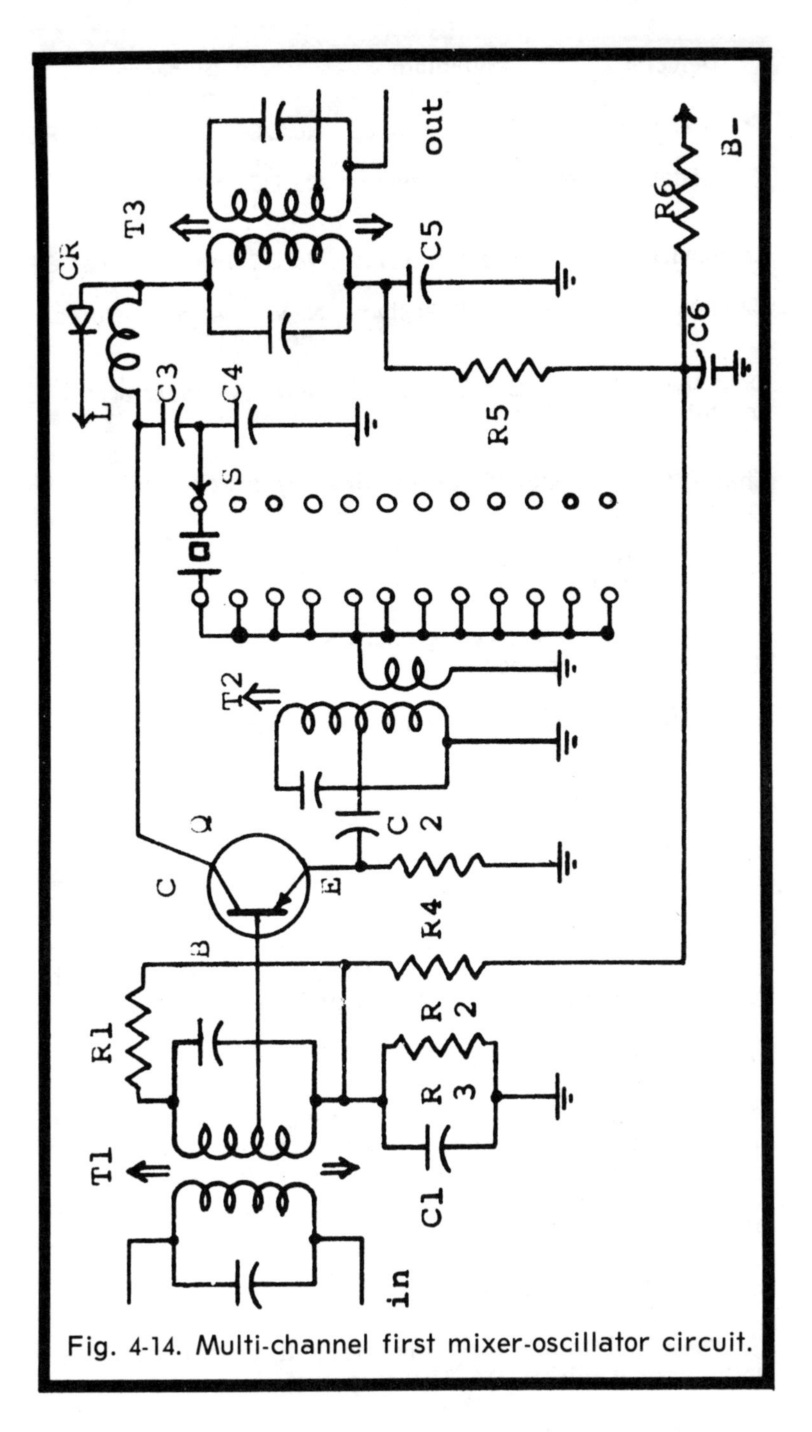

Fig. 4-14. Multi-channel first mixer-oscillator circuit.

Defect	Symptom	Evidence
Transistor open	Won't receive	No DC voltage across R4
Transistor shorted	Won't receive	No DC voltage across C-E of Q
C6 shorted	Won't receive	No DC voltage across C5
C2 open	Won't receive	New capacitor restores operation
Defective crystal	One channel out	New crystal restores operation

Chart 4-14. Troubleshooting tips, multi-channel first mixer-oscillator.

Checkout Steps

1. Measure the DC voltage across R3. Short B to E of Q. The voltage should drop.

2. Measure the DC voltage across C4. Short C1. The voltage should rise.

3. Adjust C3 for the best reception of the received signal.

455-kHz MECHANICAL FILTER

The mechanical filter shown in Fig. 4-16 is a resonator which passes signals within a very narrow frequency range centered at 455 kHz. The output of the mixer is fed through IF transformer T1 to the filter and through T2 to the IF amplifier.

Checkout Steps

1. Apply an RF test signal to the input of the mixer and measure the DC voltage at the output of the detector. Tune an RF signal generator, slowly, above and below 455 kHz. The voltage should be maximum at 455 kHz and much lower at 450 kHz and 460 kHz.

MULTI-SECTION LC IF FILTER

The signal from the first mixer is fed through T to the base of the second mixer (Q). The output of Q is fed to a low-

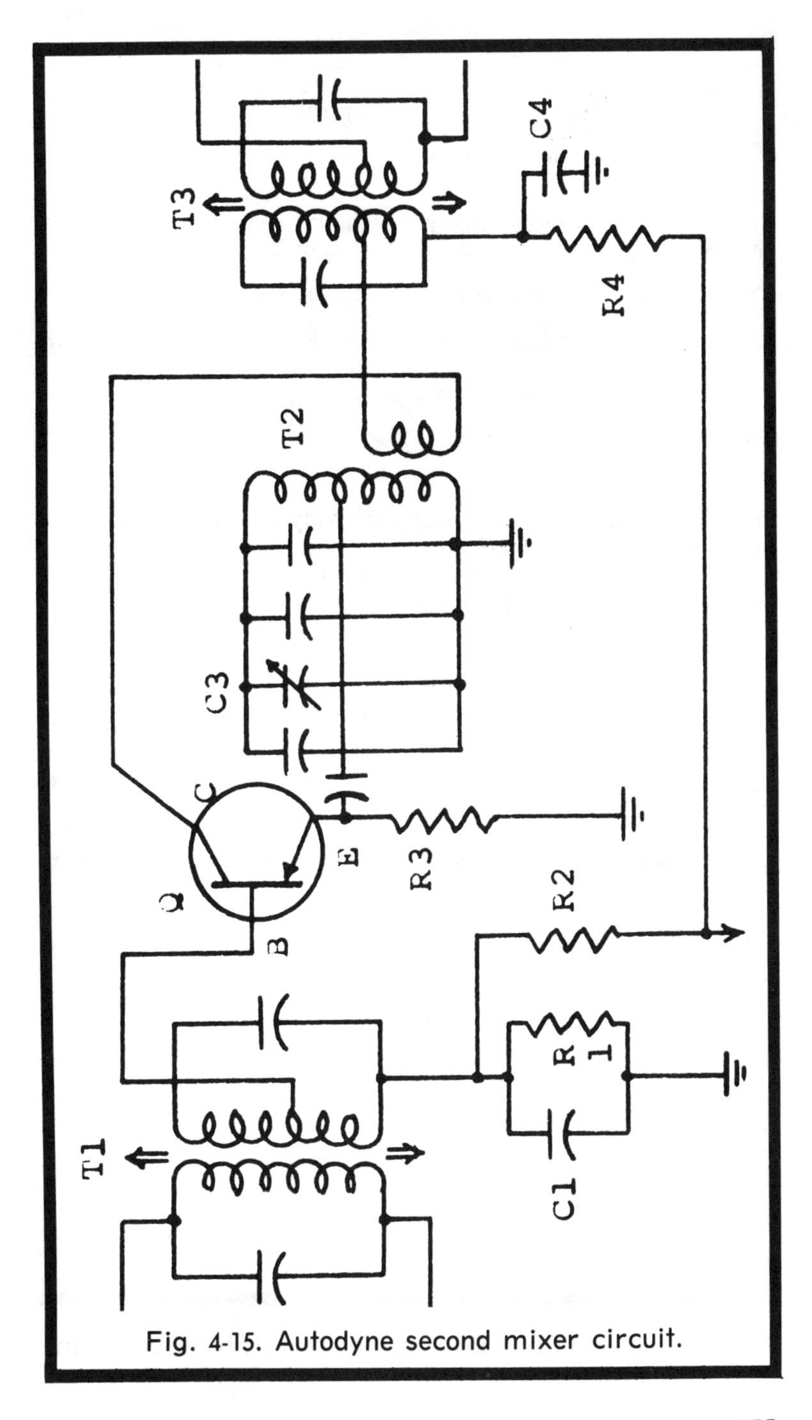

Fig. 4-15. Autodyne second mixer circuit.

Defect	Symptom	Evidence
C2 open	Won't receive	New capacitor restores operation
R3 open	Won't receive	No DC voltage across C-E of Q
Transistor shorted	Won't receive	No DC voltage across C-E of Q
Transistor open	Won't receive	No DC voltage across R3
C4 mistuned	Noisy reception	Adjusting C4 corrects problem

Chart 4-15. Second mixer troubleshooting tips.

impedance tap on L1 and is capacitively coupled to L6, then fed through C8 to the IF amplifier (not shown in Fig. 4-17). Coils L1, L2, L3, L4, L5 and L6 are carefully tuned so as to pass only a narrow band of frequencies.

Checkout Steps

1. Apply an RF test signal to the base of Q1. Measure the DC voltage at the detector output (not shown). Tune the RF signal generator slowly. The output should be maximum at the rated intermediate frequency.

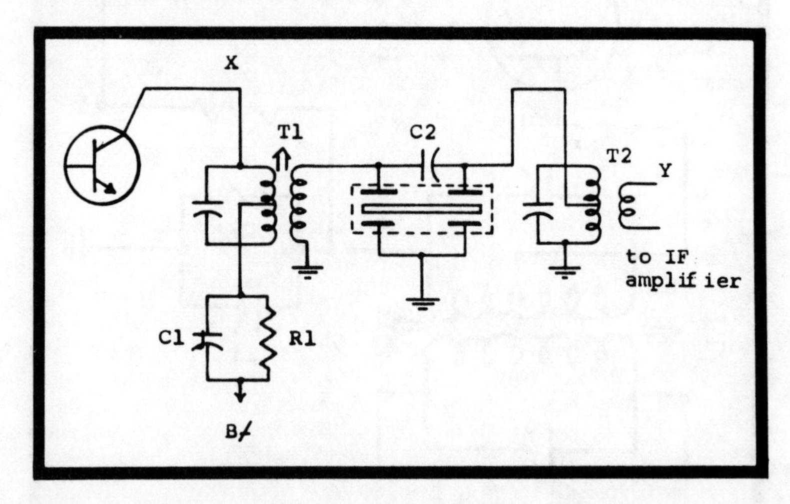

Fig. 4-16. IF amplifier 455-kHz mechanical filter diagram.

Defect	Symptom	Evidence
Defective filter	Won't receive	Receives when X connected to Y through 50-pf capacitor
T1 or T2 mistuned	Low sensitivity	Improved by tuning T1 or T2

Chart 4-16. Trouble tips, 455-kHz mechanical filter circuit.

TWO-STAGE PNP TRANSISTOR IF AMPLIFIER

The IF signal from the mixer is fed to the base of transistor Q1 through IF transformer T1, from Q1 through T2 to Q2, and through T2 to the detector (not shown in Fig. 4-18). Q1 is controlled by the AGC while the Q2 gain is fixed.

Checkout Steps

1. Apply an RF test signal to the primary of T1 and measure the DC voltage across R3. It should decrease as the test signal level increases.

2. Measure the DC voltage across R3. Momentarily short B-E of Q1. The voltage should fall.

3. Measure the DC voltage across R7. Momentarily short B-E of Q2. The voltage should fall.

DIODE DETECTOR & SERIES NOISE LIMITER

Diode CR1 (Fig. 4-19) is the detector which delivers an AF signal and a negative DC voltage to R3 and R4. R1-C1-C2 form an RF filter. The negative DC voltage across R3-R4 charges C5 through R8 and also provides the AVC voltage as indicated. Normally, diode CR2 is forward biased since its cathode is at a higher negative voltage (R3-R4) than its cathode (R4 only), and it passes AF signals to C6 and the AF amplifier. When a noise pulse is received, an instantaneous negative voltage reverse-biases CR2, opening the AF circuit. While the cathode of CR2 is also connected to the detector output through R7 and R8, C5 prevents the rise in voltage from reaching the cathode because of the time constant of R7-C5.

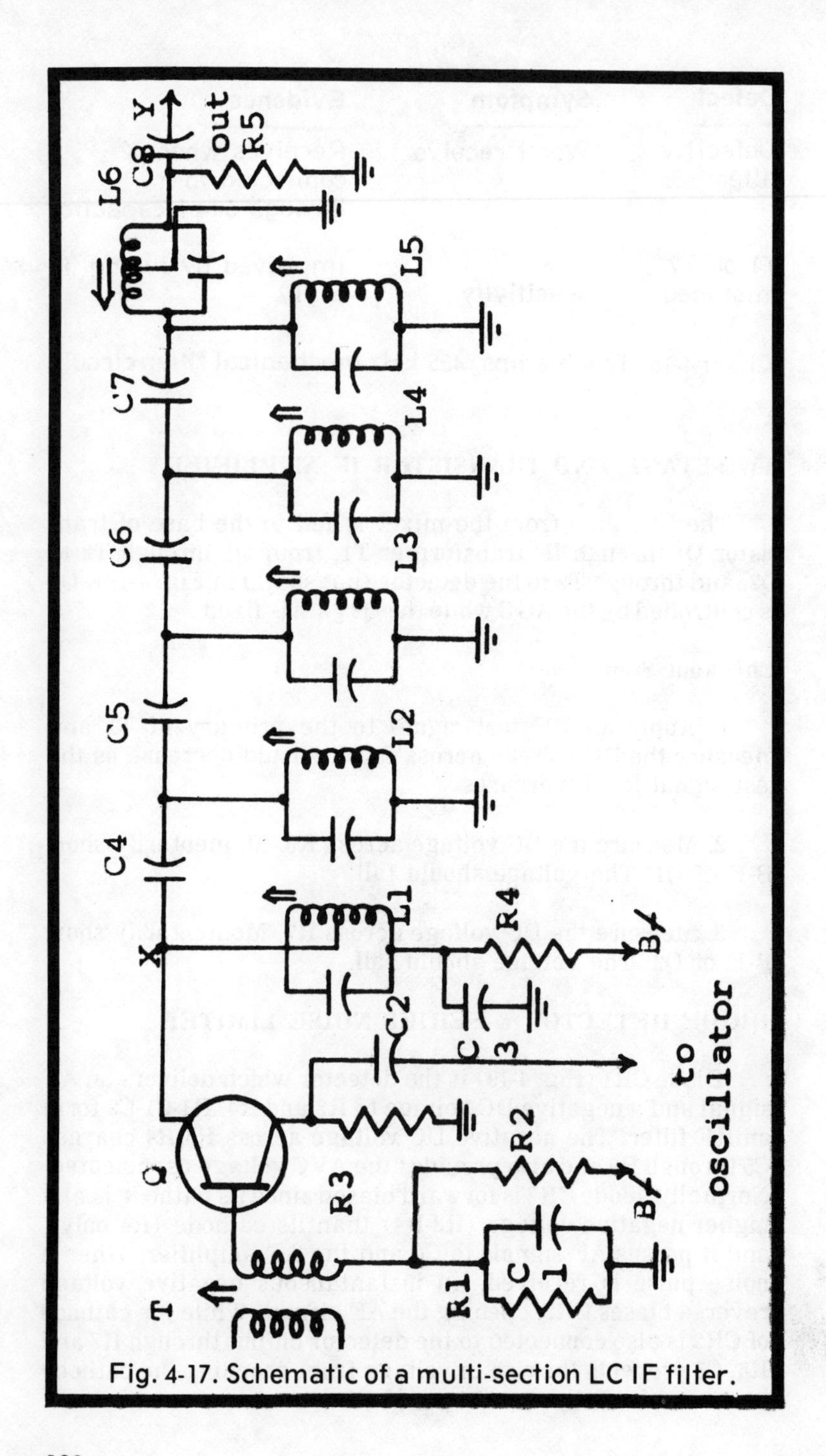

Fig. 4-17. Schematic of a multi-section LC IF filter.

Defect	Symptom	Evidence
C4, C5, C7, or C8 open	Won't receive	Receives when X is connected to Y through 50-pf capacitor
L1 open (below tap)	Won't receive	No DC voltage across R3
L2, L3, L4 or L5 open	Unselective	Adjust coil core has no effect
L6 open	Won't receive	Receives when L6 shorted

Note: Do not adjust coil cores without referring to service manual.

Chart 4-17. Multi-section IF filter trouble hints.

Checkout Steps

1. Measure the DC voltage across C5. It should rise with signal strength.

2. Same, except very momentarily short R3. The voltage should not change.

BIASED DIODE DETECTOR & SERIES NOISE LIMITER

Diode CR1 (Fig. 4-20) delivers an AF signal and a negative DC voltage which are developed across R1 and R3. The DC voltage, which varies with signal strength, is fed through R4 to the AGC circuit, through R5 to the squelch circuit, and through R6 to noise-limiter time-delay capacitor C2. Normally, noise-limiter diode CR2 is forward-biased by a positive DC voltage obtained from the junction of R2 and R3; it passes AF signals to C3 and volume control R8. When a noise pulse is received, an instantaneous negative voltage reverse biases CR2 which now acts as an open switch. Its anode is temporarily made more negative than its cathode. Although the noise pulse is also fed to its cathode through R6 and R7, the voltage is held steady by C2 because of the time constant of R6 and C2.

Checkout Steps

1. With a signal being received, measure the DC voltage across C2. It should vary with signal strength.

2. With no signal being received, measure the voltage across CR1 and CR2. A minute voltage should be present (forward bias) with the anode positive.

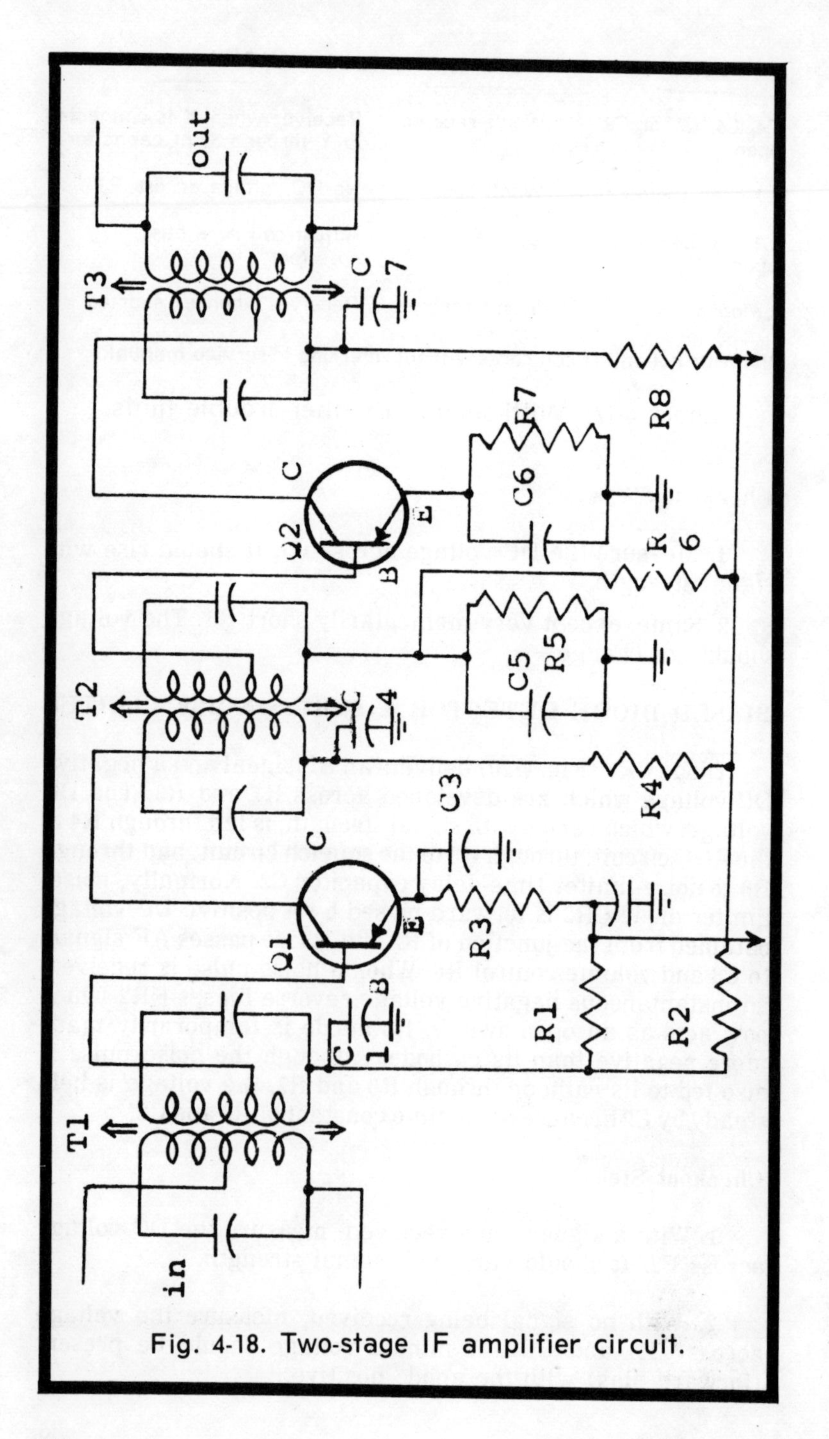

Fig. 4-18. Two-stage IF amplifier circuit.

Defect	Symptom	Evidence
Q1 open	Won't receive	No DC voltage across R3
Q1 shorted	Won't receive	No DC voltage across C-E of Q1
Q2 open	Won't receive	No DC voltage across R7
Q2 shorted	Won't receive	No DC voltage across C-E of Q2
C4 shorted	Won't receive	No DC voltage from C of Q1 to ground
C7 shorted	Won't receive	No DC voltage from C of Q2 to ground

Chart 4-18. Two-stage IF troubleshooting hints.

AMPLIFIED AGC SQUELCH WITH DIODE SWITCHING FIG. 4-21

The IF signal is demodulated by detector diode CR1. AF and a +DC voltage are developed across C1-R1. The AF

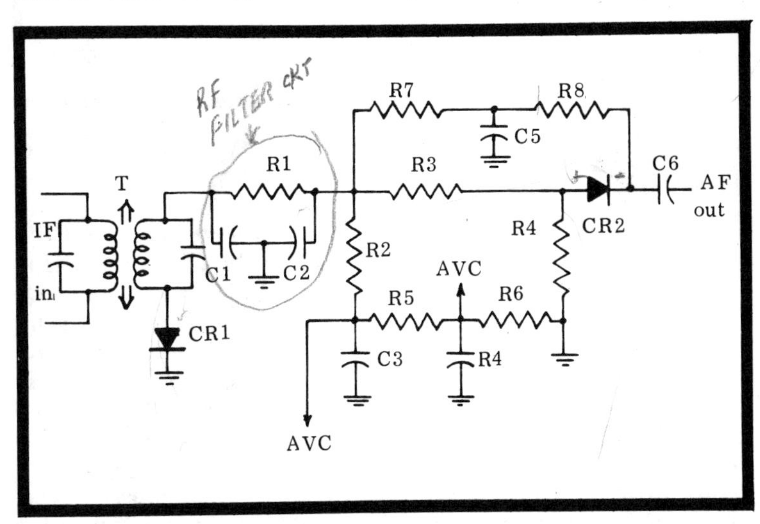

Fig. 4-19. Diode detector and series noise limiter circuit.

Defect	Symptom	Evidence
CR1 open or shorted	Won't operate	No DC voltage across C5 when signal present
CR2 open	Won't operate	Try new diode
CR2 shorted	No noise limiting	No DC voltage across CR2
C5 shorted	No noise limiting	No DC voltage across C5
C5 open	No noise limiting	Try new capacitor

Chart 4-19. Detector and noise limiter troubles.

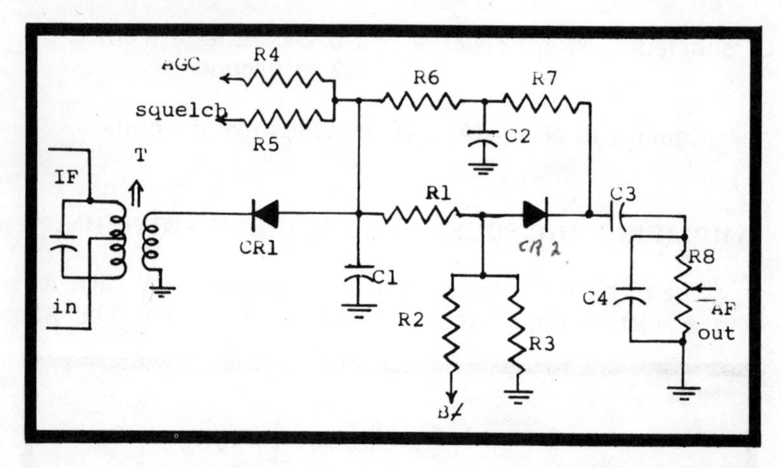

Fig. 4-20. Biased diode detector circuit with series noise limiting.

signal is fed through C8 and R16 to the input of Q3. Diode CR2, a noise limiter, shunts out noise pulses. The DC across C1-R1 is fed through R3 to the base of AGC amplifier Q1. C3 prevents Q1 from following the AF variations. When the DC rises (strong signal) the Q1 collector current rises. The voltage across R7 falls. The Q2 collector current also falls. Ordinarily, CR3 is forward-biased and acts as a closed switch; CR3 connects C5 across the AF signal path, muting the speaker. When Q2 draws less current, the voltage drop across R8 becomes lower; collector-to-emitter voltage rises, and CR3 becomes reverse-biased. CR3 acts as an open switch and C5 no longer shunts the AF signal.

Defect	Symptom	Evidence
CR1 open or shorted	Won't operate	No negative DC voltage across C1 when signal present
CR2 open	Won't receive	Try new diode
CR2 shorted	No noise limiting	No DC voltage across CR2
C2 shorted	No noise limiting	No DC voltage across C2
C2 open	No noise limiting	Try new capacitor

Chart 4-20. Biased diode detector troubleshooting tips.

Checkout Steps

1. Apply a modulated IF test signal to the anode of CR1. Measure the DC voltage at the collector of Q1. Increase the test signal level. The voltage should fall.

2. Same as above, but measure the DC voltage at the collector of Q2. DC voltage there should rise as the signal level increases.

Defect	Symptom	Evidence
CR1 open	Won't operate	No DC voltage across R1
Transistor Q1 open	Squelch won't quiet	High DC voltage across C-E of Q1
Transistor Q1 shorted	Squelch won't open	No DC voltage across C-E of Q1
Transistor Q2 open	Squelch won't quiet	High DC voltage across C-E of Q2
Transistor Q2 shorted	Squelch won't open	No DC voltage across C-E of Q2

Chart 4-21. Amplified AGC squelch troubles.

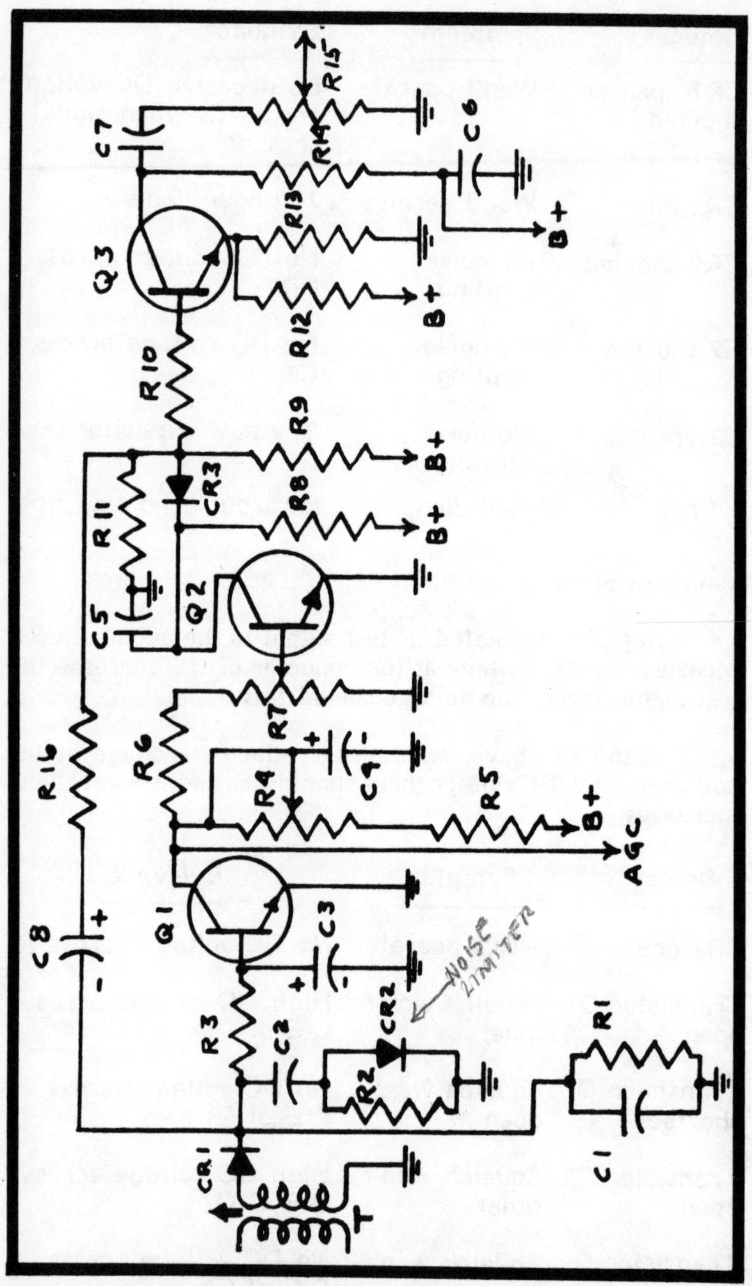

Fig. 4-21. Amplified AGC squelch circuit with diode switching.

Chapter 5

Troubleshooting Power Supply & Control Circuits

The power supply in most CB transceivers is used to supply both the transmitter and receiver with voltage. One of the first considerations, if the power supply is suspected, is to disconnect it from the rest of the unit. This will isolate the problem. Much time can be wasted checking a power supply when the actual loss of supply voltage may be caused by a shorted component in the transmitter or receiver section of the unit.

Care should be taken when measuring voltages at various points in a power supply that the meter is always on the proper scale. In tube-type equipment, higher voltages are present and the voltmeter should be on a high scale for any preliminary readings.

In many cases, when a fuse blows, it is best to first replace the fuse and test the unit again. If the fuse has been in use for a long period of time, it may have just opened due to fatigue. There is nothing seriously wrong with the power supply, except that the fuse should be replaced.

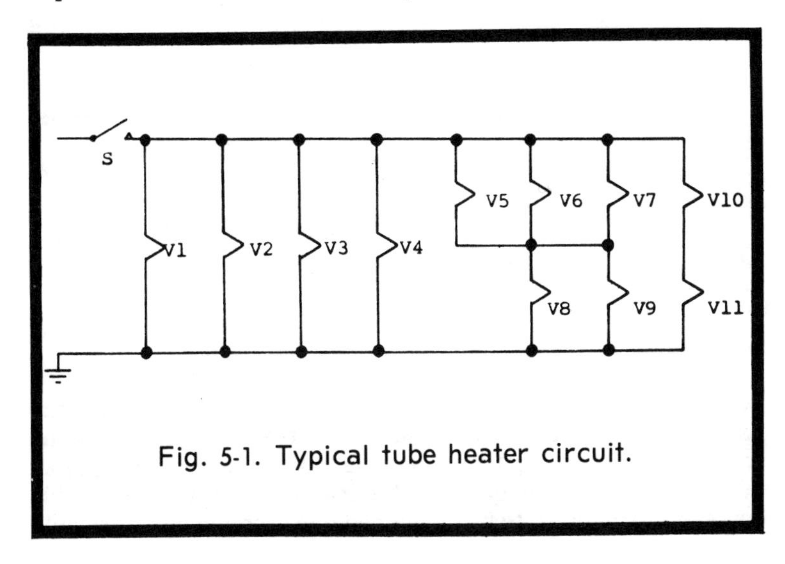

Fig. 5-1. Typical tube heater circuit.

Defect	Symptoms	How to Check
V10 or V11 open	Neither tube lights	Try new tubes
V8 or V9 open	V5, V6, V7 light dimly	Try new tubes
V5, V6 or V7 open	V8, V9 light dimly	Try new tubes
V1, V2, V3 or V4 open	Defective tube does not light	Try new tube
S open	Tubes don't light	Short circuit switch

Chart 5-1. Tube heater circuit trouble tips.

TYPICAL TUBE HEATER CIRCUITS

Six-volt tube heaters are connected in series-parallel (V5, V6, V7, V8, V9 in Fig. 5-1) and in series (V10, V11) across a 12-volt supply. 12-volt tubes are connected in parallel (V1, V2, V3, V4).

Checkout Steps

1. Measure the voltage across S when it is open. It should read 12 volts or more. When S is closed, the voltage should be zero.

2. Measure the voltage across V1 with S closed. It should read 12-14 volts.

3. Measure the voltage across V5. It should read 6-7 volts.

4. Measure the voltage across V10; it should read 6-7 volts.

Note: Use an AC voltmeter for base station checks and a DC voltmeter for a mobile unit.

VIBRATOR POWER SUPPLY

In Fig. 5-2 DC from the battery is fed through L, which with C1, C2 and C3 forms a hash filter to the vibrator and the tapped low-voltage primary of power transformer T. As the vibrator contacts flip from one side to the other, DC is applied alternately to each half of the transformer winding. This causes a form of AC to be developed and stepped up to 100 volts or more at the secondary. This AC voltage is rectified by diodes CR1 and CR2 connected in a voltage doubler circuit.

For AC operation, AC power is fed to the other primary winding and the vibrator does not operate. Tube heater voltage is obtained from the tertiary winding of the low-voltage primary. For DC operation, tube heaters are fed directly from the DC source.

Checkout Steps

1. Measure the DC voltage across C10. It should be higher than normal at first and then drop as the tubes warm up.

2. Turn the power on and off to determine that the vibrator starts every time.

Note: When replacing the vibrator, also replace C6.

TRANSISTOR SWITCHING POWER SUPPLY

Power transistors Q1 and Q2 in Fig. 5-3 operate as a push-pull oscillator and as electronic switches. When Q1 is conducting, Q2 is cut off, and vice versa (see vibrator). Collector current flows through the tapped primary of T2. Transformer T1 is the feedback transformer which supports oscillation. Note: The 117-volt AC input winding is used only for AC operation.

Checkout Steps

1. Measure the AC voltage across the secondary of T2 with 12.6 volts DC input. Reduce the input to 11 volts. The AC voltage should still be present.

2. Measure the DC voltage from C of Q1 to C of Q2. It should be zero or almost.

SOLID-STATE MOBILE UNIT POWER SUPPLY

Battery voltage is fed to the "+" and "-" terminals in Fig. 5-4. If polarity is accidentally reversed, diode CR conducts and the fuse blows. When S (on-off switch) is closed, lamp I glows and B- power is fed to the transistors. The positive power lead goes to the floating common ground through S. Capacitors C1 and C2 and choke L reject ignition noise from the battery circuit and provide RF and AF paths to chassis ground. C3 functions as an AF bypass.

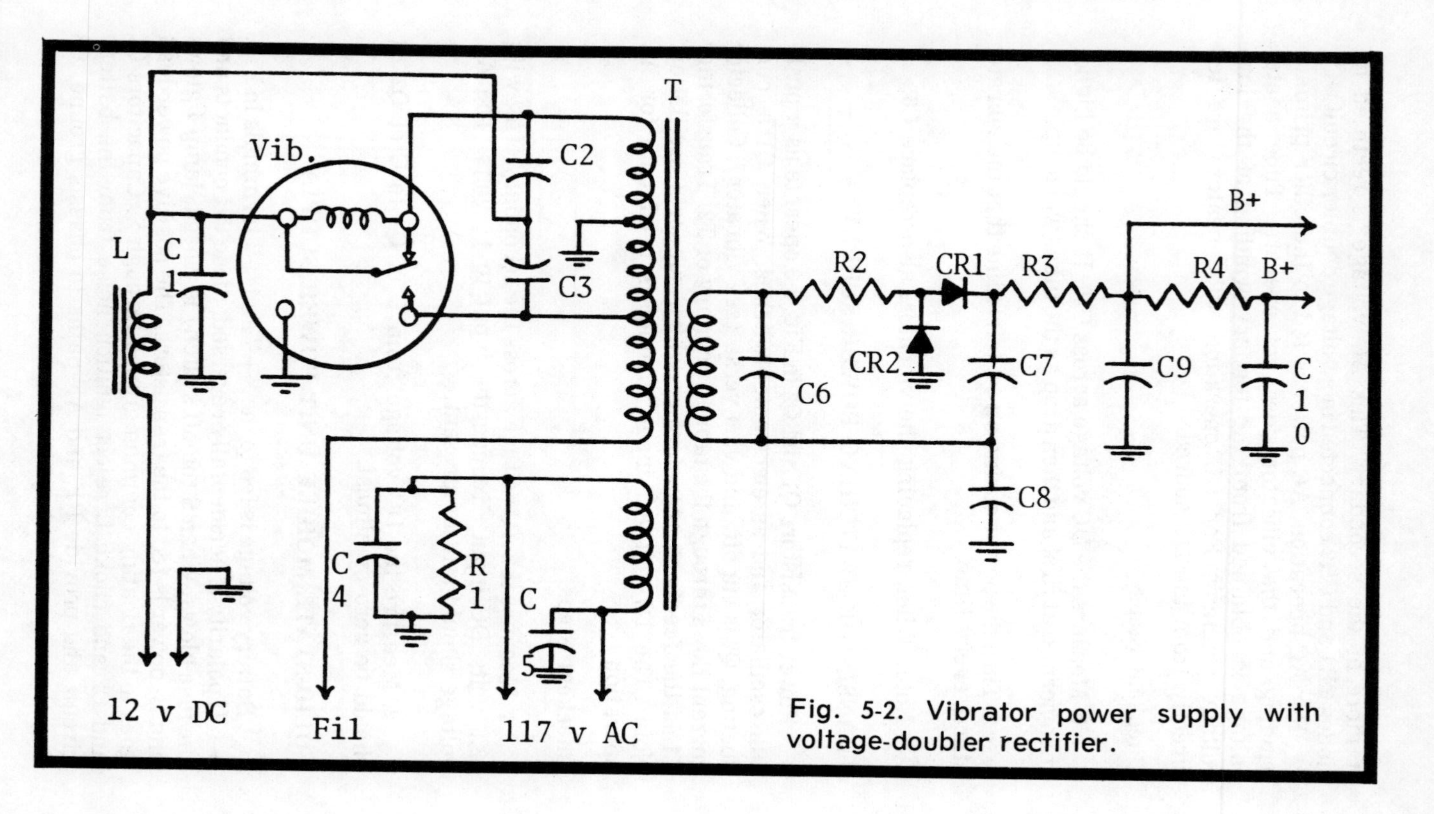

Fig. 5-2. Vibrator power supply with voltage-doubler rectifier.

Defect	Symptom	Evidence
Worn out vibrator	No DC output voltage	New vibrator restores operation
Open rectifier, or C7 or C8 shorted or open	Low DC output voltage	No DC voltage across C7 or C8
C6 shorted or leaky	No or low DC output voltage	Low or no AC voltage across C6
C9 or C10 shorted	No DC output voltage	No DC voltage across C9 or C10
C7 or C8 shorted	Low DC output voltage	No DC voltage across C7 or C8

Chart 5-2. Vibrator power supply trouble tips.

Checkout Steps

1. Measure the DC voltage across C3 with S closed. The lamp should light and 12-14 volts should be indicated.

2. Measure the resistance between chassis and common ground. The meter should read zero with S open or closed.

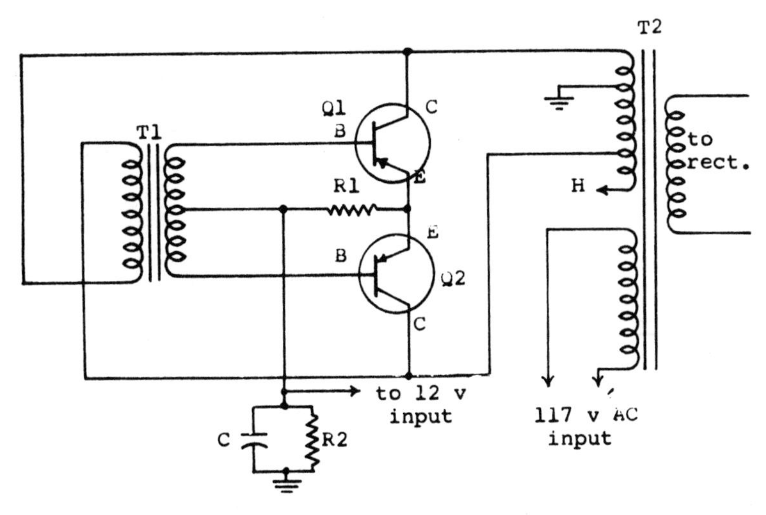

Fig. 5-3. Transistor switching power supply for tube-type transceivers.

Defect	Symptom	Evidence
Q1 shorted	Won't operate	No DC voltage across C-E of Q1
Q2 shorted	Won't operate	No DC voltage across C-E of Q2
T1 primary open	Won't operate	No AC voltage across T2 secondary
T2 secondary open	Won't operate	No DC voltage across B-E of Q1 or Q2
R1 open	Won't operate	No DC voltage across E of Q1 and Q2 and ground

Chart 5-3. Troubleshooting hints for a transistor switching power supply.

SHUNT ZENER REGULATED POWER SUPPLY

Full battery voltage is applied to the lamp and some circuits when S is closed. If the battery polarity is accidentally reversed, CR1 does not conduct and damage to transistors is prevented. Resistor R drops voltage which is regulated by zener diode CR2. C1 and C2 bypass RF and AF to ground. Relay contacts (K) apply voltage to the transmitter or receiver.

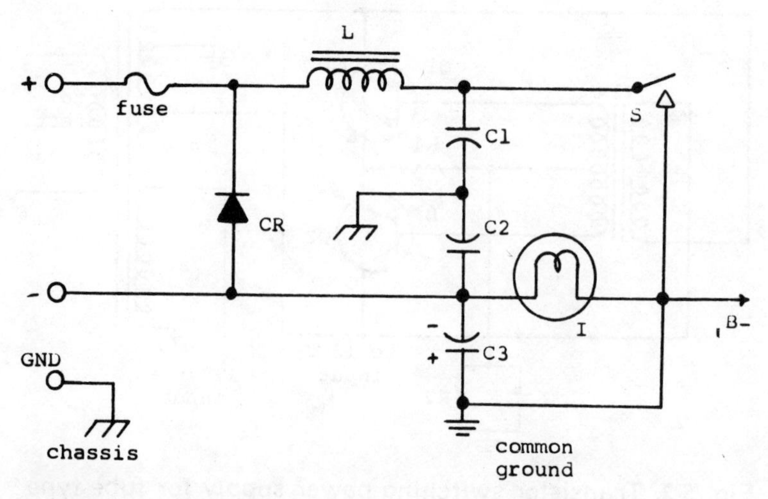

Fig. 5-4. Typical mobile unit power supply circuit.

Defect	Symptom	How to Check
Fuse open	Lamp does not light	Try new fuse
Lamp open	Lamp does not light	Try new lamp
Choke open	Lamp does not light	Short circuit choke
Switch open	Lamp does not light	Short circuit switch
C1, C2 or C3 open	Noisy reception with engine running	Shunt new capacitor across C1, C2 or C3
C3 shorted	Lamp does not light	Measure voltage across C3

Chart 5-4. Troubles occurring in a typical solid-state mobile unit power supply.

Checkout Steps

1. Measure the DC voltage across CR2 with the unit set to receive and when set to transmit. The voltage should not vary.

2. Measure the DC voltage across R. The voltage should vary somewhat when the unit is switched from receive to transmit.

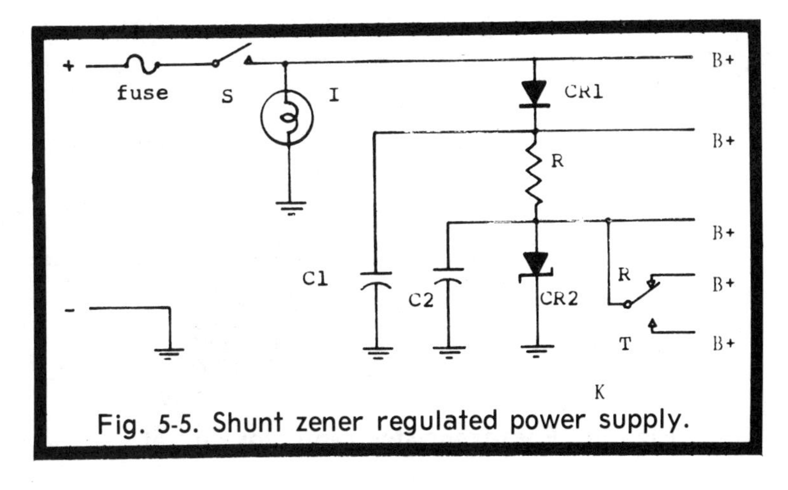

Fig. 5-5. Shunt zener regulated power supply.

Defect	Symptom	How to Check
Fuse open	Lamp does not light	Try new fuse
Switch open	Lamp does not light	Short circuit switch
CR1 open	Transceiver does not operate	Measure voltage across CR1. Should be almost zero. Measure voltage across R. Should be about 2 volts.

Chart 5-5. Zener regulated power supply troubles.

PASS-TRANSISTOR REGULATOR CIRCUIT

Full battery voltage is fed directly to some circuits and through L to others, and through the contacts of relay K. When K is in the receive position (R), regulated voltage is fed to some receiver transistors. Transistor Q functions as a pass-transistor whose resistance rises as the input voltage rises, and vice versa. Zener diode CR provides a reference voltage to the base of the transistor.

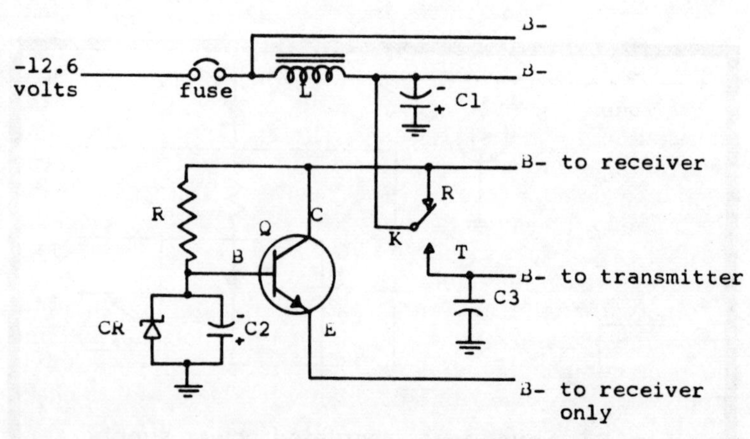

Fig. 5-6. Pass-transistor regulator circuit.

Defect	Symptom	Evidence
Fuse open	No operation	OK with new fuse
L open	No operation but TR relay clicks	No DC voltage across C1
C1 shorted	Fuse blows	No DC voltage across C1
C3 shorted	Fuse blows	No DC voltage across C3 (in transmit)
Transistor open	Won't receive	No DC voltage across C2
Transistor shorted	Unstable receiver	No DC voltage across C-E of Q
R open	Won't receive	No DC voltage across C2
C2 shorted	Won't receive	No DC voltage across C2
CR shorted	Won't receive	No DC voltage across C2

Chart 5-6. Trouble tips, pass transistor regulator.

Checkout Steps

1. Measure the DC voltage across E of Q and common ground. It should be 9-11 volts and present only in the receive position.
2. Measure the DC voltage across C3. It should be 12-14 volts in the transmit position.

CONTROL CIRCUITS

The most common trouble in a control circuit is the malfunctioning of the TR relay. The contacts may become pitted or dirty and as a result there will be no circuit between the antenna and the transceiver or operating voltages may not be supplied properly to the unit. Relays should be replaced if possible rather than repaired unless there is an experienced relay repairman available.

Another common fault is the microphone push-to-talk button and its associated wiring. If the relay does not pull into the transmit position, the switch may not be operating to supply the relay coil with DC voltage. It is also well to check all the external plugs and jacks to make sure they are mechanically correct.

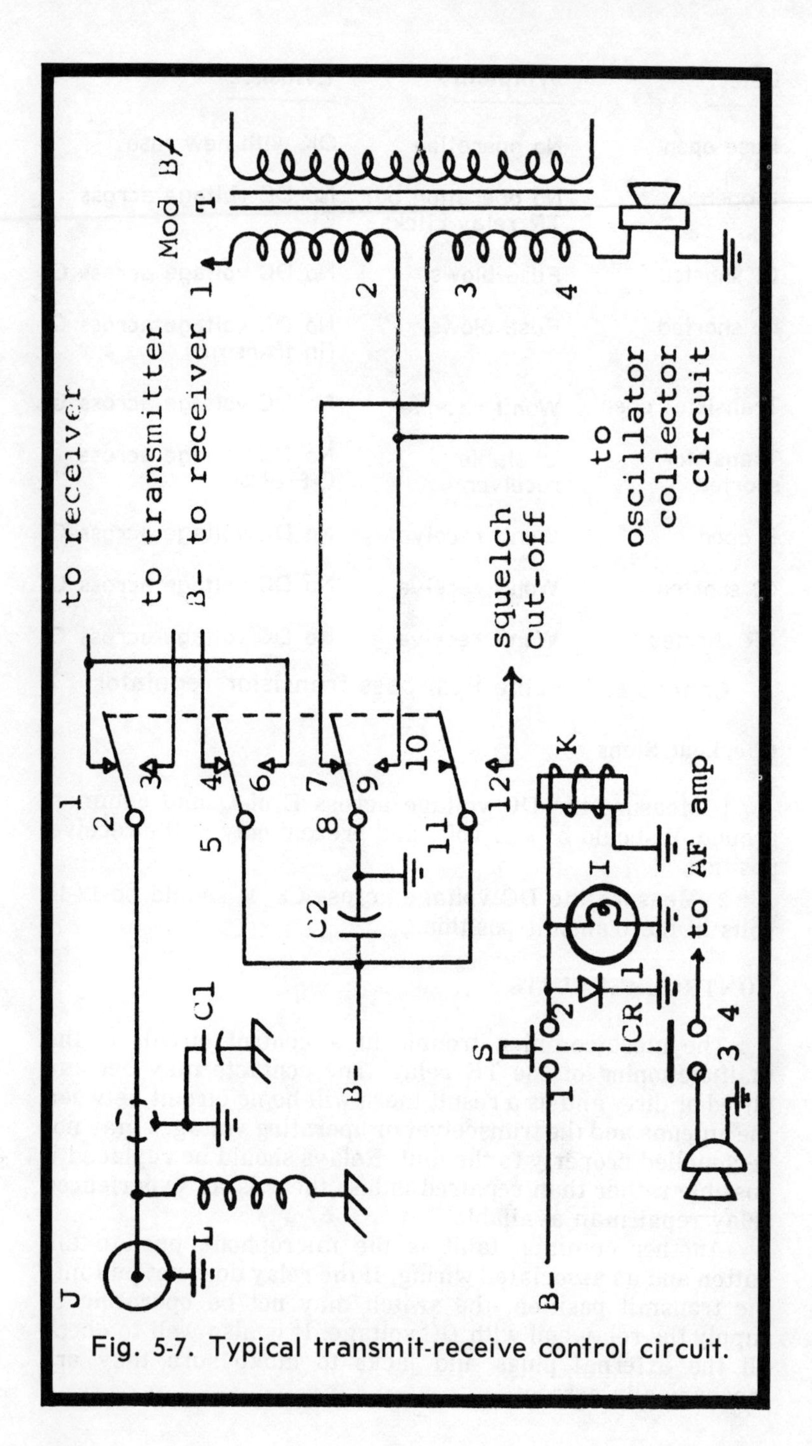

Fig. 5-7. Typical transmit-receive control circuit.

Defect	Symptom	How to Check
Defective mike switch or cord	Won't transmit	Connect anode of CR to B-. Relay should pull in.
Open relay coil	Won't transmit	Same as above.
Shorted diode	Won't transmit	Disconnect diode.
Relay contact defects	Won't transmit	Short contacts 8-9, 11-12 in that order.
Relay contact defects	Won't receive	Short contacts 7-8, 4-5, 1-2 in that order.
Relay contact defects	Won't transmit	Short 1 and 2 of T. Transmitter should operate but not modulate.

Chart 5-7. Troubles frequently found in control circuits.

TYPICAL TRANSMIT-RECEIVE CONTROL CIRCUIT

When receiving, contacts 1 and 2 of relay K in Fig. 5-7 connect the receiver input to antenna jack J. Contacts 4 and 5 apply DC voltage to the receiver. Contacts 7 and 8 complete the speaker circuit through ground. To transmit, the mike PTT pushbutton S is closed. Contacts 1 and 2 apply DC voltage to pilot lamp I and the coil of relay K, which pulls in. Contacts 3 and 4 complete the microphone circuit. Relay contacts 2 and 3 connect the transmitter output to antenna jack J. Contacts 5 and 6 disable the receiver. Contacts 8 and 9 ground terminal 2 of modulation transformer T, completing the DC voltage path to the RF power amplifier. Also, the transmit oscillator is activated. Contacts 11 and 12 disable the squelch so the AF amplifier will operate. When S is released, diode CR acts as a surge protector.

Checkout Steps

1. Close S. The lamp should light and the relay should pull in.

2. While talking into the mike with S closed, ground contact 7 of S. Voice should be heard through the speaker.

3. Measure the RF output at J. Power should be present when S is closed and should rise when talking into the mike.

4. Release S. The receiver should operate.

Defect	Symptom	Evidence
		Shorting contacts noted restores operation:
Defective switch	Won't receive	1 and 2 of S
	PA won't work	Same when grounding 4
	Won't transmit	5 and 6 (transmit switch on)
External speaker jack	Won't receive	1 and 2 of J2
Relay contacts	Won't receive	2 and 3 of K
	Won't transmit	2 and 3 of K (transmit switch on)
	PA won't work	2 and 3 of K (S in PA position)

Chart 5-8. Troubles associated with the audio output control circuit.

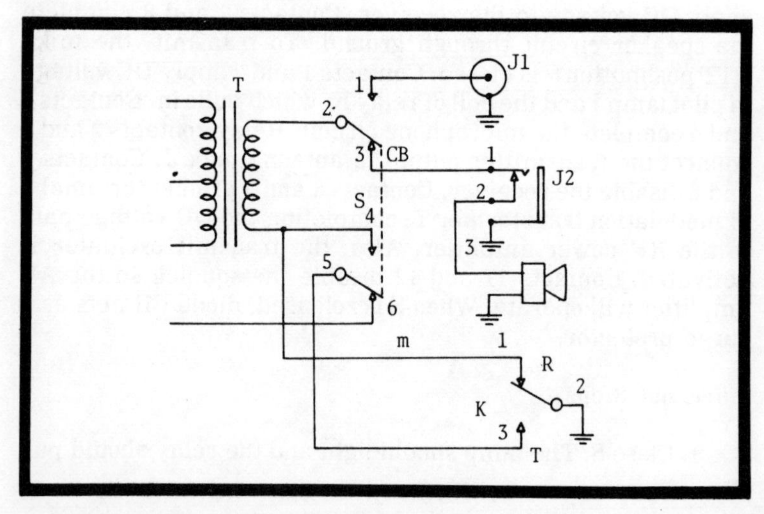

Fig. 5-8. Audio output control circuit with public address and external speaker jacks.

AUDIO OUTPUT CONTROL WITH PUBLIC ADDRESS AND EXTERNAL SPEAKER JACKS

Audio is fed, when receiving, from output transformer T through contacts 2 and 3 of PA-CB switch S and contacts of an external speaker (or headphone) jack J2 to the speaker (Fig. 5-8). Contacts 1 and 2 of T-R relay K ground one side of secondary of T to complete the audio path through contacts 5 and 6 of S. When transmitting, relay contacts 1 and 2 open, cutting off the speaker. Contacts 2 and 3 energize the transmitter. When using the public address feature, S is set to the PA position. Audio from T is now fed through contacts 1 and 2 of S to J1. Contacts 5 and 6 of S open and prevent contacts 2 and 3 of K from energizing the transmitter. Contacts 4 and 5 of S complete the PA speaker circuit through contacts 2 and 3 of K.

Checkout Steps

1. With the receiver on and the squelch wide open, listen for noise in the speaker with S in the CB position. Operate the transmitter; noise should cease.

2. Connect an external speaker to J2. Noise should be heard through it (receiver on) but not through the internal speaker.

3. Set S in the PA position and connect an external speaker to J1. Noise should not be heard with the receiver on. Turn the transmitter on and talk into the mike. Voice should be heard through the external speaker.

Chapter 6

Modifications FCC & EIA Standards

A CB transmitter must perform in accordance with the technical standards specified in Part 95, FCC Rules and Regulations, and in Canada as specified in the regulations of the Department of Communications for the General Radio Service. A summary of these standards is included in this section.

It is also well to be familiar with the EIA (Electronic Industries Association) standards for AM transceivers used in the Citizens Radio Service. These standards specify test procedures and serve the public interest through the elimination of any misunderstanding as to what the conditions were when such tests were made.

No modifications (see the last section in this chapter) to a CB transmitter may be made except by or under the supervision of a licensed first-class or second-class radiotelephone operator. If any modification affects the transmitter so that it no longer operates in accordance with the technical standards of the FCC (in the United States) or of the DOC (in Canada), the FCC license is no longer valid.

Therefore, it is suggested that no circuit changes be made unless authorized by the manufacturer who is then liable to the FCC for any deviation from the published specifications.

(1) Channels 1 through 8 and 10 through 23 (see Table 6-1) may be used between units of the same station.

(2) Only Channels 10, 11, 12, 13, 14, 15 and 23 may be used for communications between units of different stations.

(3) Channel 9 is the emergency channel and may be used only for communications involving the immediate safety of life of individuals or the protection of property or any communication necessary to render assistance to a motorist.

Priorities have been assigned by the FCC in regard to the use of Channel 9 and they are:

Channel	Frequency (MHz)
1	26.965
2	26.975
3	26.985
4	27.005
5	27.015
6	27.025
7	27.035
8	27.055
9	27.065
10	27.075
11	27.085
12	27.105
13	27.115
14	27.125
15	27.135
16	27.155
17	27.165
18	27.175
19	27.185
20	27.205
21	27.215
22	27.225
23	27.255

Table 6-1. List of the CB channel frequencies.

(1) Any message relating to an existing situation dangerous to life or property; i.e., fire, automobile accident, etc.

(2) Any message relating to a hazardous situation; i.e., car stalled in a dangerous place, lost child, boat out of gas, etc.

(3) Road assistance to a disabled vehicle on the highway or street.

(4) Road and street directions.

FCC TECHNICAL STANDARDS

The technical standards as set forth by the FCC for stations in the Citizens Radio Service under Class D licenses are:

1) CB stations may use radiotelephony only.

2) The average power input to the final plate or collector circuit may not exceed 5 watts.

3) The average power supplied to the antenna system may not exceed 4 watts. (This measurement shall be made at the output of the transmitter.)

4) The frequency tolerance of the transmitter shall be maintained within 0.005 percent of the authorized frequency.

5) Stations in this service are authorized to use amplitude voice modulation, including single sideband and-or reduced or suppressed carrier.

6) Tone signals or signaling devices used to activate receivers is permitted; however, using tones to attract attention or for remote control is prohibited.

7) The maximum bandwidth allowed in Class D service is 8 kHz.

8) On any frequency removed from the assigned frequency by more than 50 percent, and up to and including 100 percent of the authorized bandwidth, the emissions must be attenuated by at least 25 db. From 100 percent to 250 percent of the assigned frequency the attenuation must be at least 35 db

and on any frequency greater than 250 percent, the attenuation must be greater than 50 db.

9) When using amplitude modulation in this service, the modulation shall not exceed 100 percent on either positive or negative peaks.

10) The nameplate should always be attached to the transmitter in a prominent place, showing the type or model number, serial number, manufacturer's name, class of station and any FCC data that may be required.

11) Any controls that may effect changes in the carrier frequency should not be accessible from the exterior.

EIA STANDARDS

1. Standard power supply test voltages:

Nominal Voltage	Test Voltage
6v DC	6.6v
12v DC	13.6v
24v DC	26.4v
32v DC	36.0v
64v DC	72.0v
110v DC	110.0v
110-220v AC	117.0v

2. Standard temperature and humidity:

20 degrees C to 30 degrees C 0 to 90 percent humidity

30 degrees C to 35 degrees C 0 to 70 percent humidity

3. Standard output load:

50-ohm resistive termination.

4. Standard test receiver audio response:

Plus or minus 1 dB from 300 to 3000 Hz.

5. Standard input signal source:

A calibrated signal generator whose terminal resistance is 50 ohms.

6. Standard test input signal:

It shall be modulated at 1000 Hz at such a level to produce 30 percent modulation.

7. Minimum standard:

Over a test voltage range as specified in No. 1, with a variation of plus or minus 10 percent, the limits shall be:

a) Transmitter power output: 2 db
b) Receiver audio output: 2 db
c) Usable sensitivity: 2 db
d) Squelch threshold: 2 db
e) Transmitter stability maintained over a supply voltage range of plus or minus 15 percent.
f) Transmitter power shall not drop more than 3 db at a test voltage -20 percent.
g) Receiver sensitivity shall be at least 1.0 microvolt for a 10 db signal-to-noise ratio.
h) Adjacent-channel selectivity shall be at least 30 db.
i) Spurious response attenuation shall be at least 25 db, including IF rejection, except the image which shall be 10 db.

When a CB transceiver is either modified or repaired, it is well to supply the owner with a check sheet when it is returned to service. This gives him a positive indication that his unit is operating legally and the check list may be shown to the FCC field engineer in the event that any question might arise. This is the equivalent of the log book entries that are required in other radio services and it gives the user much more confidence in his maintenance technician. As an adjunct to this test report, a lable should be attached to the unit verifying the latest test date.

Many citations and fines have been levied by the monitoring division of the FCC for off-frequency operation recently and it is suggested that these tests—particularly the

OPERATIONAL CHECK

MANUFACTURER: ____________________

MODEL NUMBER: ____________________

SERIAL NUMBER: ____________________

OWNER: ____________________

STATION CALL SIGN: ____________________

DATE OF TESTS: ____________________

POWER OUTPUT INTO 50-OHM LOAD ____________________

POWER INPUT (FINAL RF STAGE ____________________

FREQUENCY MEASURED:

CH 1	______ MHz	CH 12	______ MHz
CH 2	______ MHz	CH 13	______ MHz
CH 3	______ MHz	CH 14	______ MHz
CH 4	______ MHz	CH 15	______ MHz
CH 5	______ MHz	CH 16	______ MHz
CH 6	______ MHz	CH 17	______ MHz
CH 7	______ MHz	CH 18	______ MHz
CH 8	______ MHz	CH 19	______ MHz
CH 9	______ MHz	CH 20	______ MHz
CH 10	______ MHz	CH 21	______ MHz
CH 11	______ MHz	CH 22	______ MHz
		CH 23	______ MHz

MODULATION CHECK: ____________________

RECEIVER SENSITIVITY: ____________________

SIGNATURE OF TECHNICIAN

GRADE OF LICENSE ____________________

DATE OF EXPIRATION ____________________

Fig. 6-1. This form, or a similar one, should be used by a technician to verify the results of various tests.

frequency check—be carried out at certain intervals as is done in other services. An example of a test sheet report is given in Fig. 6-1. It may be used as a sample by the maintenance technician.

FIELD MODIFICATIONS

As mentioned at the beginning of this chapter, no modifications may be made that will affect the operation of a transmitter. However, there are several changes that may be made without official sanction.

Microphone

One of the most common modifications is to replace the microphone with either a better quality one or with a standard handset. The major problems are the choice of impedance and the gain of the new unit. If the new microphone has a different impedance, a matching transformer should be used. These may be purchased at any local supplier and come in a wide variety of impedances. Care should be exercised in adding anything in a low-level audio stage so that noise and hum pick-up are not increased. Grounds should be carefully chosen, and if the transformer is mounted on the chassis it should also have a solid mechanical connection. If the microphone has three leads, the shield on the audio lead is used as a common ground for both audio and the push-to-talk circuit. If there is a separate ground return for the push-to-talk switch, it may be jumpered to the audio shield at the microphone connector.

When the gain of a new microphone is different from the gain of the original microphone, it is well to install a variable resistor in the input line of the transmitter to prevent audio distortion and overmodulation. A resistor of 50,000 ohms will not change the audio characteristics of the circuit but will be adequate to compensate for the change in gain.

Some newer microphones have a transistor preamplifier built into the housing. If one of these is added to the transceiver, a low-voltage source must be used to supply the transistor with power. This simply means adding another lead to the microphone cable and connecting it to the proper voltage source within the transceiver.

If the operator desires privacy when using his equipment, a handset should be employed. The switch on the handset hanger will cut off the audio to the speaker when the handset is removed. The necessary changes to replace a standard microphone with a handset and hanger are shown in Fig. 6-2.

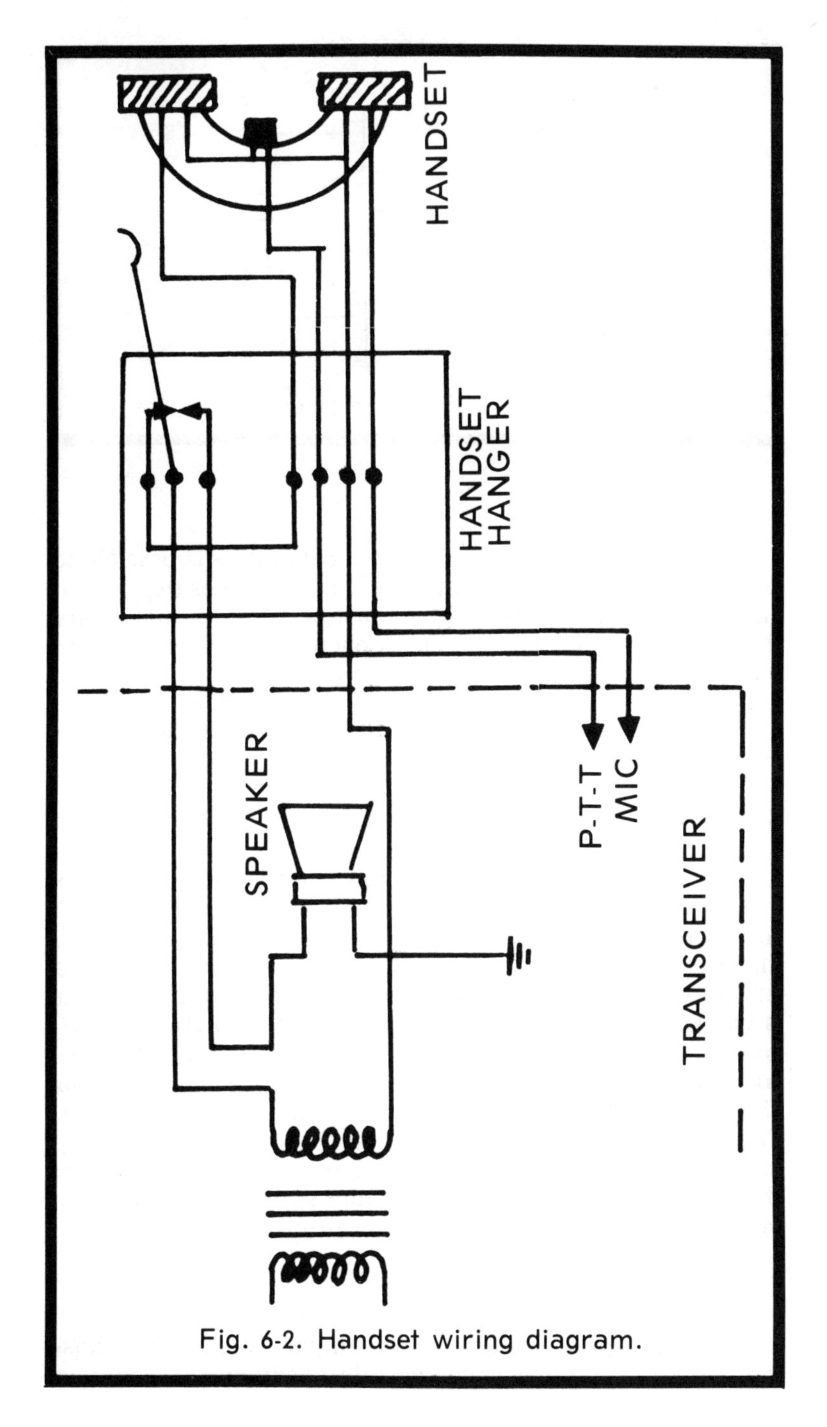

Fig. 6-2. Handset wiring diagram.

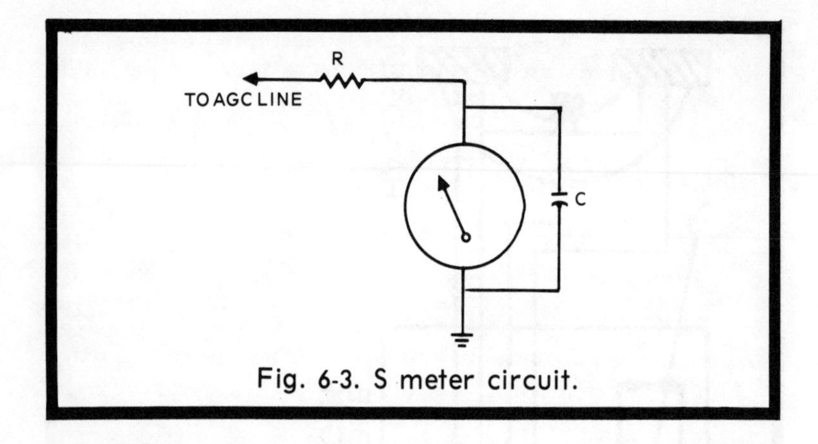

Fig. 6-3. S meter circuit.

Adding Channels

If the transceiver is a single-channel unit, is is possible to add more frequencies by installing a multiple-position switch and extra crystal sockets. Care must be taken to make sure that the circuit inductance and capacity are not changed so that the crystals will operate on the proper frequency. If the unit has been type approved or type accepted by the FCC, modification **cannot** be made unless new data is submitted, proving that the unit still meets the published specifications.

S Meter

Many of the more expensive transceivers have a built-in S meter, which gives an indication of the strength of a radio signal that is being received. This is a very simple addition that requires only three components, a 50 microammeter, a 0.25-mfd capacitor and a small resistor. The metering circuit is connected after the detector at the point where the AGC voltage is fed back to the RF or IF amplifiers. Fig. 6-3 shows how the circuit should be connected. The value of the resistor (R) is determined by the available voltage on the AGC line. The value of this resistor should be chosen so that a strong incoming signal gives a full-scale deflection of the meter. It should be of a sufficiently high value so as not to upset the operation of the detector and AGC circuits.

Filters

Several types of filters are available to improve transmission and reception under certain conditions. A common complaint is the noise that is present on the power lines at a base station site. Several manufacturers make these filters and they are simply inserted between the source of the AC power and the supply line to the transceiver. This same filter also will eliminate any RF signals that are fed back into the power line from the transmitter, causing television interference.

There are three basic filters that can be installed in the RF or antenna circuit. Suppliers may have special names for them, but the results are the same. When the transceiver is found to be very inefficient because of a mismatch between the antenna and the output circuit of the transceiver, a tunable filter may be installed so that the standing wave ratio (SWR) is reduced and the proper 50-ohm impedance is reflected back to the transmitter for more efficient transmission.

Another type of filter, commonly called a wave trap, is designed to eliminate a specific frequency either from being received or from being transmitted. This filter is most efficient when it is mounted as close as possible to the transceiver antenna connector and is well grounded at that point. This type filter is usually the most inexpensive way to eliminate any problems that arise with interference on TV Channel 2. A more elaborate filter is referred to as a "low-pass" filter. This eliminates any transmission and reception above a specific frequency, usually about 30 MHz. It is well to know the amount of rejection that this type of filter provides before it is installed.

One of the most frequently heard complaints in CB operation is the ignition noise and other sharp impulse noises that interfere with reception in certain areas of poor reception. A new accessory has recently been marketed that is quite effective and also very easy to install in the field. It is installed in the antenna circuit and consists of a preamplifier and noise eliminator. The impulses are amplified and blanking pulses are mixed with them to neutralize them. The preamplifier stages also give the receiver added sensitivity and selectivity so that the overall receiver performance is greatly improved.

Tone Alert

Many CB units have an accessory that is commonly called "selective calling." This is a system of tone-signaling which allows the receiver to remain squelched until a specific signal is received. No "on-channel" RF signal will be heard if it does not transmit the correct tone that the receiver unit is designed to receive.

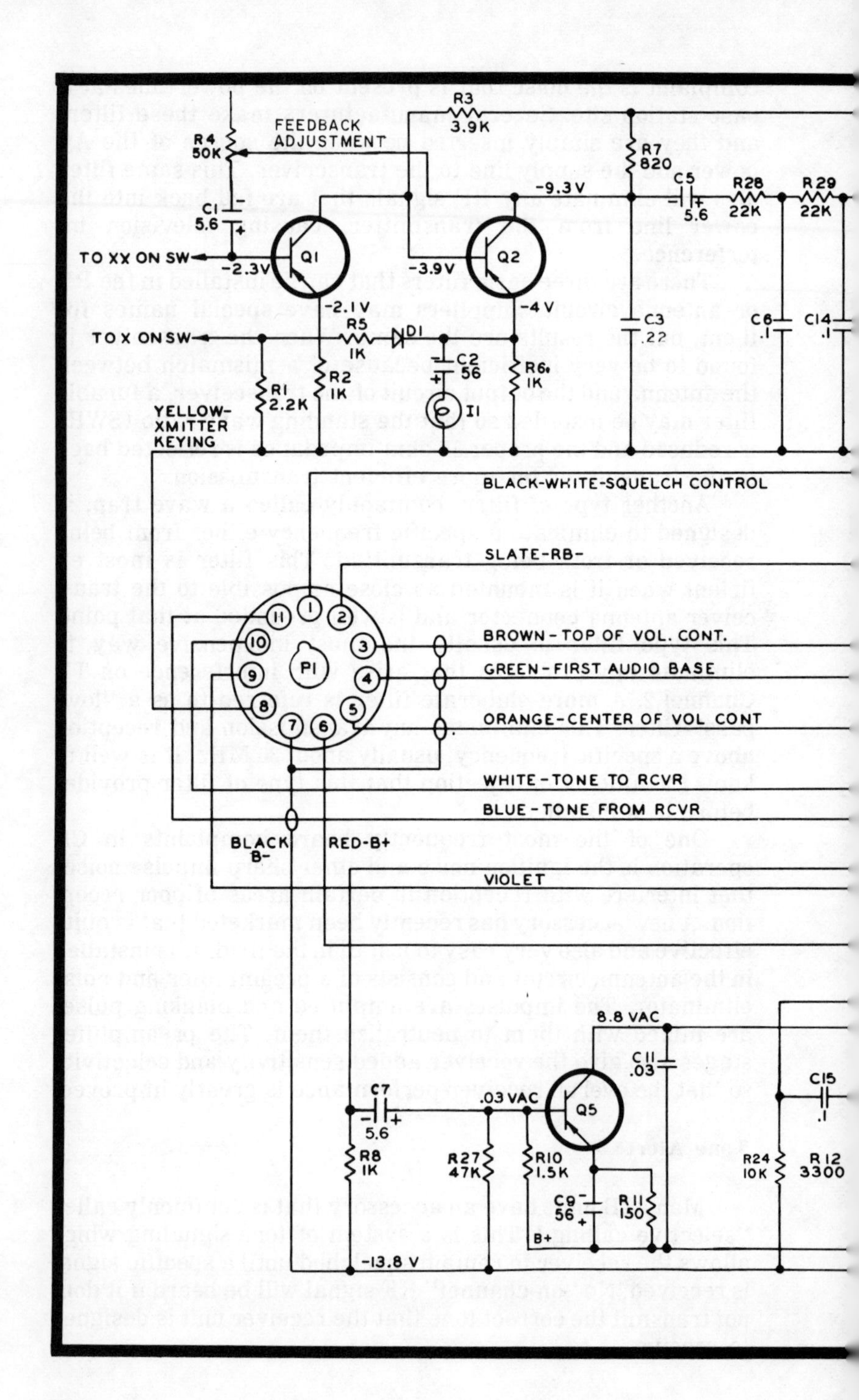
R3
3.9K
R4
50K
FEEDBACK
ADJUSTMENT
R7
820
-9.3 V
C5
5.6
R28
22K
R29
22K
C1
5.6
TO XX ON SW
-2.3V
Q1
-3.9V
Q2
-2.1 V
-4 V
R5
1K
D1
TO X ON SW
C3
.22
C6
.1
C14
.1
C2
56
R2
1K
R6
1K
R1
2.2K
I1
YELLOW-
XMITTER
KEYING
BLACK-WHITE-SQUELCH CONTROL
SLATE-RB-
BROWN-TOP OF VOL. CONT.
GREEN-FIRST AUDIO BASE
ORANGE-CENTER OF VOL CONT
WHITE-TONE TO RCVR
BLUE-TONE FROM RCVR
P1
BLACK
B-
RED-B+
VIOLET
8.8 VAC
C11
.03
C15
.1
C7
5.6
.03 VAC
Q5
R8
1K
R27
47K
R10
1.5K
R24
10K
R12
3300
C9
56
R11
150
B+
-13.8 V

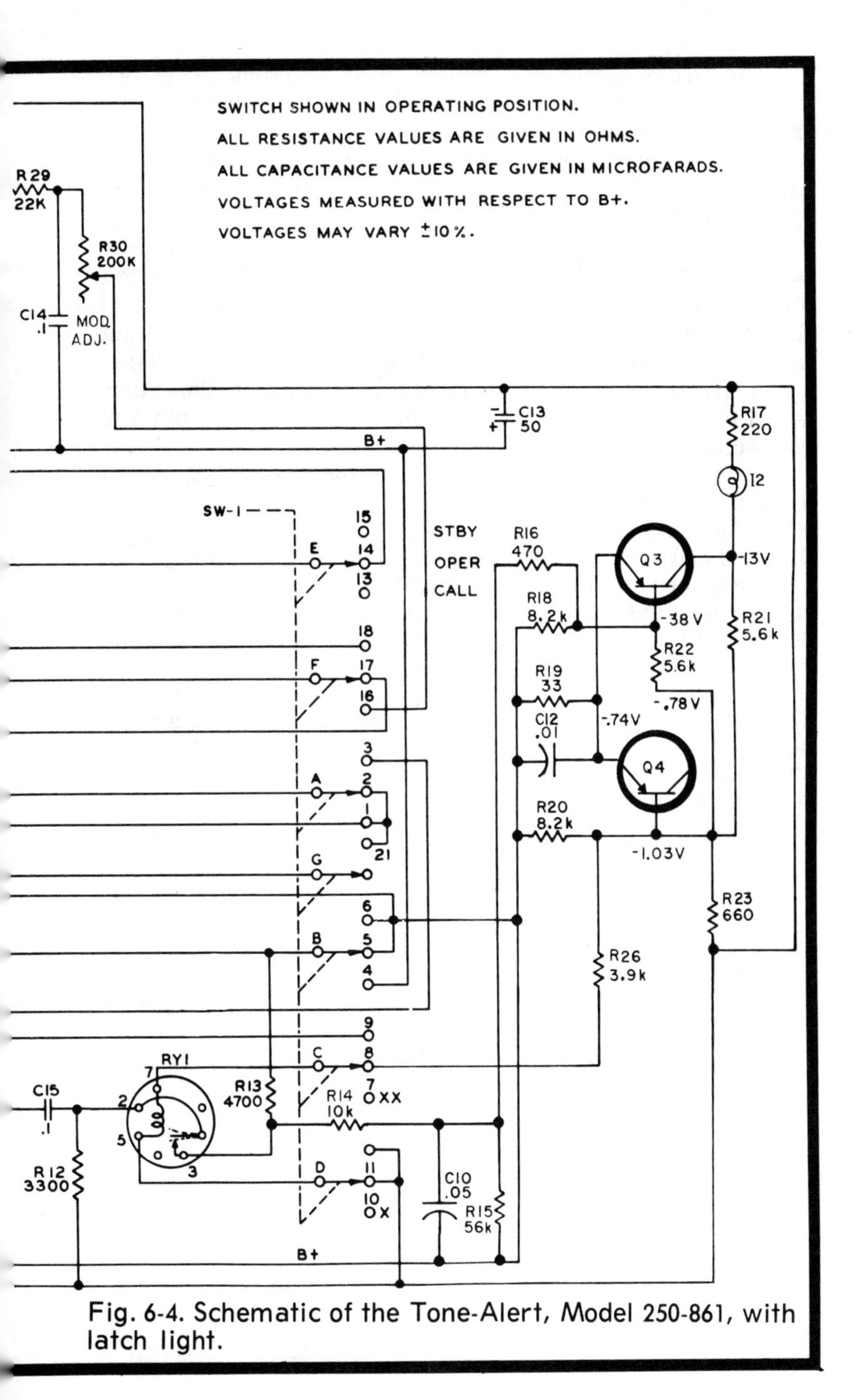

Fig. 6-4. Schematic of the Tone-Alert, Model 250-861, with latch light.

The "Tone-Alert" is a widely used signaling accessory that may be attached to any standard CB transceiver. The operation is as follows (see Fig. 6-4):

When the operator moves the function switch (SW-1) to the call position, the audio oscillator (Q1, Q2) is placed in operation. The frequency of oscillation is determined by the tone reed (RY1). Control of the oscillator frequency and the rapid start of oscillation is controlled by a feedback circuit and the amount of feedback is determined by the setting of control R4. The network (C3, C4, C6, C15, R28, R29) shapes the output to give a good sine wave. The modulation adjustment (R30) is set to give 90 percent modulation of the transmitter. This audio is fed to the first audio stage of the transmitter via pin 4 of connector P1.

In the standby position, the incoming tone is connected through the connector (P1) to the base of the audio amplifier (Q5). The collector of this transistor is placed in series with the reed coil and when the proper tone is received a negative pulse appears on the base of Q3, half of a multivibrator. This causes the transistor to conduct and the second stage (Q4) to become cut off. The front panel light (I2) is illuminated and remains on until the unit is reset. This is accomplished by moving the switch to the operate position to receive the call. The incoming tone also is heard on the receiver to alert the operator if he does not see the panel light come on. When the switch is in the operate position, all calls are heard on the receiver whether the proper tone is present or not.

Chapter 7

Typical Transceiver Circuits

This chapter contains a brief analysis of some typical CB transceivers. These include relatively simple hand-held units and also the more complex units that include SSB capabilities. In most instances, the basic circuitry is the same and only the added features of each unit change.

In most units any variation from the standard tuning and test procedures are detailed on the schematic or in the operating instructions of the transceiver. It should be noted again that an FCC license is required to tune the transmitter section of a transceiver and the input power to the final amplifier must not exceed 5 watts.

COURIER MODEL CCT4

The Courier Model CCT4 (see foldout Panel A) is one of the more sophisticated hand-held type CB transceivers. It has a full 5-watt input to the final stage of the transmitter and is capable of operation on all 23 of the CB channels. This unit incorporates most of the features to be found in a base station or mobile unit. It is equipped with a meter that indicates battery condition, RF power and received signal strength ("S"). It has delta tuning, public address output, squelch and jacks for external speaker, microphone and antenna. The power supply on the back of the transceiver may be interchanged with a rechargeable ni-cad pack or an AC power supply if it is used as a base station.

The CCT4 contains 16 transistors, one FET, one integrated circuit, 14 diodes and 14 crystals. The 14 crystals are employed in a synthesizer type circuit that actually employes three separate oscillators. A high-frequency oscillator that contains six crystals is used for both transmit and receive frequencies to heterodyne with either the four crystals in the transmitter oscillator or the four crystals in the receiver oscillator to obtain the correct on-channel frequency.

In the transmitter the two oscillator frequencies are combined in the mixer stage and then fed to the buffer amplifier. This signal is increased to the required level in the driver stage to feed the RF amplifier. The signal is then passed through the pi network to the antenna. A small portion of the signal is fed through a dropping resistor and rectified by diode D503 to give the RF power reading on the meter. To tune the transmitter use an RF wattmeter and a 50-ohm dummy load in place of the antenna and simply tune L905, L906, L907, L908, L910 and L911 for a maximum output reading. Care must be taken, however, in not exceeding the 5-watt input limitation to the final power amplifier, Q907. Coil L912 should be tuned with a field strength meter to give maximum output when using the whip antenna on the unit.

The receiver section of the Model CCT4 has an RF amplifier (Q101) which is protected from overload by limiter diodes D101 and D102. The on-channel signal is mixed with the master oscillator signal in the first mixer stage (Q301) and fed to the second mixer stage (Q302) where it is heterodyned with the receiver oscillator output to produce the 455-kHz IF signal. The first 455-kHz IF amplifier stage has a ceramic filter to give better selectivity and prevent adjacent-channel interference. Diode detector D305 feeds the AGC signal back to the RF amplifier and the second mixer and is regulated by two separate diodes.

Diode D501 acts as a noise limiter and transistors Q502 and Q503 are the squelch amplifiers. The first audio amplifier and driver stages are combined in an integrated circuit to conserve space. The audio output is a push-pull amplifier that may also be used as a public address amplifier when an external speaker is connected to jack J1.

DYNASCAN COBRA 25

The Cobra 25 (Fig. 7-1) has a total of 14 crystals in a synthesizer circuit designed to operate on all 23 of the CB channels. The master oscillator contains six crystals in the 17-MHz range that heterodyne with the transmitter crystals (10-MHz range) to produce the on-channel signal. (See foldout Panel B for a complete schematic.) The voltage supplied to the oscillator stages and the transmitter is regulated by zener diodes at the critical points and the capacitors used in these circuits are temperature compensated for additional stability. The resultant RF signal is filtered through a triple-tuned circuit to suppress any unwanted harmonics as well as the fundamental crystal frequencies. Further filtering to prevent radiation from the shielded oscillator section is provided by L5 in the feeder connection to the transmitter section.

The modulator section has a low-pass filter (C50, L4 and C51) between the preamplifier and driver stages to prevent

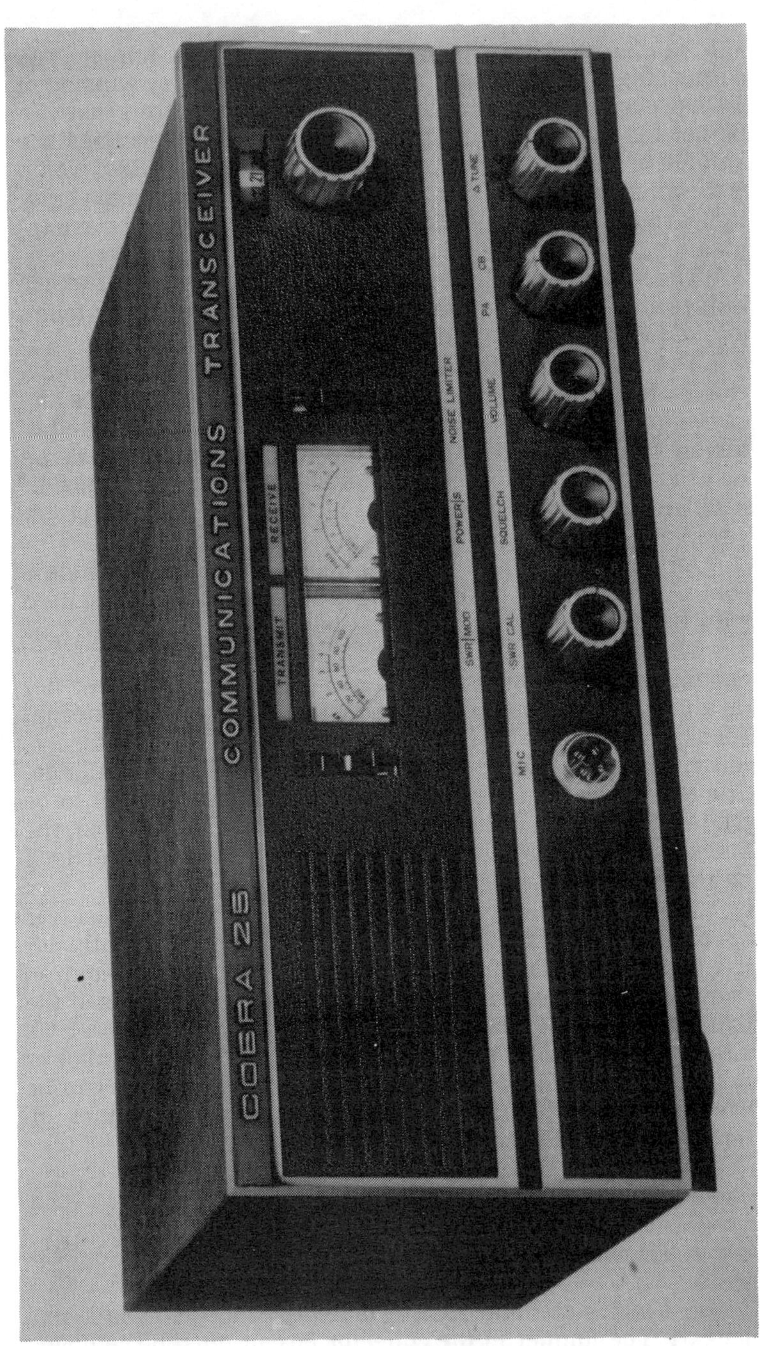

Fig. 7-1. Cobra Model 25 transceiver.

any frequencies above 2700 Hz from being amplified. The output stages have an AGC circuit in the secondary winding of T9 to prevent overmodulation. There are two controls that can be set for this function. One is VR6, which determines the amount of feedback supplied and the other one is VR206 which is the microphone gain control used to regulate the audio input for various microphones that could be used with the equipment.

The final RF amplifier has a modified pi circuit to further reduce any harmonic radiations and also to match the antenna and cable properly to the transmitter.

The receiver section of the transceiver employs double-conversion, and an FET in the first mixer stage minimizes any cross-modulation problems. The local oscillator uses "delta" tuning so that any variation of the incoming signal may be cleared up. The change is approximately plus or minus 1.7 kHz. An integrated circuit is used in the IF section of the receiver for more efficiency.

An additional feature of this unit is the microphone gain control which allows different types of microphones to be used with the modulation capabilities unchanged.

This unit is also equipped with a "public address" capability which means that the RF sections of the unit are disabled and the audio section is used to feed an external speaker to provide 3 watts of audio power. This feature is incorporated in many of the CB transceivers now available. The PA-CB switch (Fig. 7-2) allows the receiver output to be fed to a PA system. When the switch is in the PA location, the audio transformers output is switched to the PA speaker jack on the rear panel.

The delta-tune feature can greatly enhance receiver operation. For example, if a received signal is slightly off frequency, the delta tune control can be operated as required to optimize the received signal level. The effectiveness of the delta-tune feature under these conditions can be observed either by listening for a more readable signal at the speaker or by noting the S-meter reading. The delta-tune control can be used also to minimize or eliminate adjacent-channel interference by an additional 10 to 20 db.

Tuning Procedure

1. Transmitter

a) A wattmeter and dummy load are connected to antenna jack J1. The jumper in the collector circuit must be removed and a milliammeter inserted in the circuit.

b) Key the transmitter and check for any indication of power output.

c) The following coils are then tuned for a maximum output: T101, T102, T103, T104, T10, T11, L8, L9 and L10.

d) With 13.8 volts on the collector of Q16, the current reading on the meter in the collector must not exceed 360 ma. (This complies with the maximum input to the final amplifier of 5 watts.)

e) If the current exceeds 360 ma, L8 and L9 should be unloaded until the correct current value is obtained at a maximum power output value.

f) The modulator section of the transmitter must be adjusted to give the proper level for the microphone being used. This control is set to give a good signal without over-modulating. If an oscilloscope is available, the waveform should be checked for distortion.

2. **Receiver**

a) Connect a signal generator to antenna jack J1 and an audio voltmeter across the speaker leads or plug the meter into J3 (if J3 is used, an external 8-ohm loading resistor should also be used).

b) With the signal generator modulated internally at 1000-Hz, a low signal should show an indication on the meter when the generator is tuned to the proper channel. If no signal is

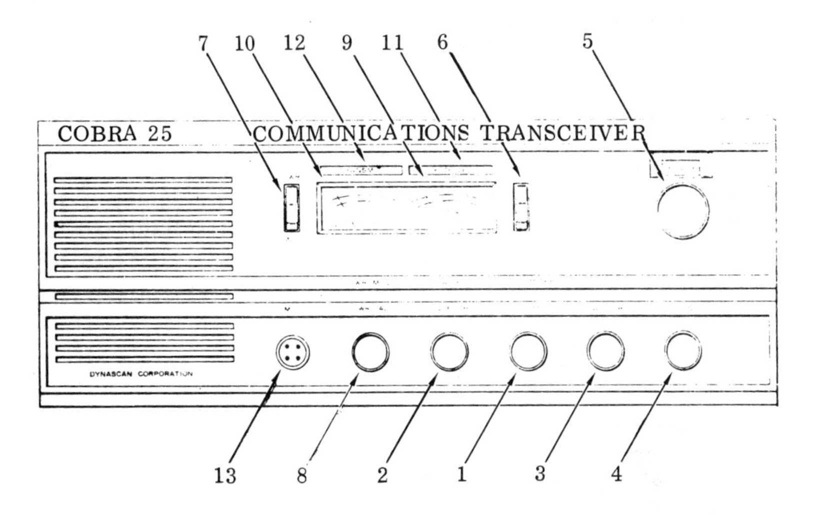

Fig. 7-2. Front panel control functions.

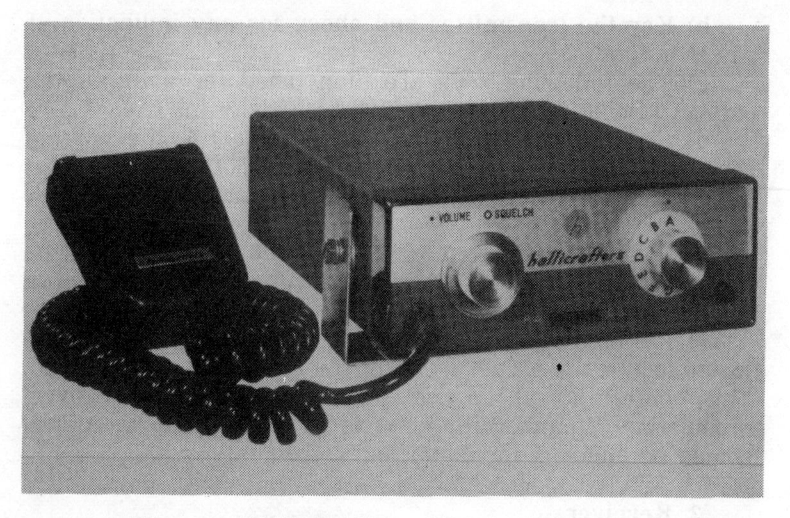

Fig. 7-3. Hallicrafters CB-21 transceiver.

obtained even with a stronger output from the generator, the generator input should be moved to T4 and returned to 455 kHz. T4, T5, T6 and T7 should be tuned for a maximum reading on the voltmeter. The generator input should then be moved to the base of Q2 and tuned to 10 MHz. T2 is tuned for a maximum reading on the meter.

c) Steps a and b are now repeated and all of the coils are checked to give the maximum output reading on the meter for a minimum signal from the generator. The signal output of the generator should be no more than 1.0 microvolt.

HALLICRAFTERS MODEL CB-21

The Hallicrafters Model CB-21 is an 8-channel mobile unit that is fully solid-state (Fig. 7-3). It has a complement of 17 transistors, five diodes and two zener voltage regulators. The unit is designed to operate from the standard 12.6-volt automobile supply. Specifications for the unit are:

Transmitter:

Power input: 5 watts
Power output: 3.2 watts
TVI suppression: Second harmonic trap
Crystal type: CR-81-U third overtone, plus or minus .005 percent tolerance.
Antenna impedance: 50 ohms

Receiver

Sensitivity: 1.0 microvolt for 10 db SNR
Audio output: 3.5 watts

Chassis Removal

To remove the chassis from the cabinet, take off the mounting handle. Then remove the four screws on the bottom, holding the chassis to the cabinet, and carefully slide the chassis out from the front.

Alignment

All alignment and performance specifications stipulated were performed at the EIA standard DC input of 13.8 volts. For receiver alignment the following equipment is required (see Fig. 7-4).

1. Standard AM-type signal generator covering the frequency range of at least 455 kHz to 27.255 MHz, modulated 30 percent with either 400 or 1000 Hz. Generator should be capable of being accurately adjusted to 1650 kHz.
2. Output meter (or AC vacuum tube voltmeter) connected across the speaker terminals (or 8.0-ohm termination).
3. 0.1 mfd 200v capacitor.

For transmitter alignment:

1. 50-ohm non-reactive dummy load (two 100-ohm 2-watt resistors in parallel).
2. RF power output indicator connected across above the load.
3. 0-500 milliampere DC meter.

Transmitter alignment: The transmitter oscillator coil, T1, has been adjusted at the factory for series-resonant crystal operation. This coil should not be tampered with, as off-frequency illegal operation may result. The FCC requires that persons making transmitter frequency adjustments be licensed commercial radio-telephone operators, second class or higher, and that they have adequate frequency-measuring equipment. For proper on-frequency operation of this transceiver, use only standard military-type CR81-U, third-overtone, series-resonant crystals.

Coil L3 and capacitor C16 function as a trap circuit to suppress second harmonic radiation. Unless a receiver tuning the 54-55 MHz range and if an S-meter is available, the setting of C16 should not be changed. If such a receiver is available, the receiver should be tuned to the second harmonic of the CB-

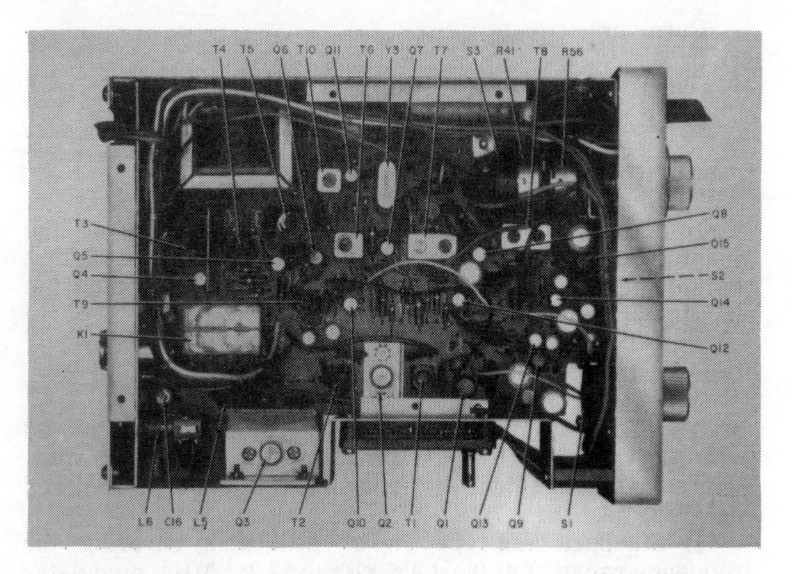

Fig. 7-4. Component locations, Hallicrafters CB-21.

21 (2 x operating frequency). With the CB-21 in transmit, C16 should be adjusted for minimum indication on the receiver.

JOHNSON MESSENGER III

Fig. 7-5 is a block diagram of the Messenger III. Access to the chassis is gained by removing the four screws at the rear of the cabinet. Place the transceiver in an upside down position on a flat surface. Grasp the front panel and carefully withdraw the chassis assembly from the cabinet. A complete schematic appears on foldout Panel D.

Receiver Alignment & Troubleshooting

A. Equipment Required:

Power supply: 13.8v DC regulated, 1 amp minimum.
Oscilloscope
AC VTVM
RF generator: .455 to 30 MHz with an attenuated output of 1 microvolt to .1 volt, capable of modulation at 1000 Hz at 30 percent.
Audio signal generator: 1000 Hz
VTVM with RF probe
6 db pad connected to the output of the signal generator for troubleshooting and alignment.
B. Test equipment setup (Fig. 7-6).

Alignment	Connections	Generator Frequency	Channel Crystal	Adjust
455 kHz IF transformer	Signal generator to 2nd mixer base through 0.1-mfd capacitor.	455 kHz plus or minus 0.2 percent	None	Top of T6, the top and bottom of T7 and T8. Keep reducing the generator output to maintain the output level below ½ watt (volume control fully clockwise).
1650-kHz IF transformer	Signal generator to 1st mixer base through 0.1-mfd capacitor.	Tune for peak at 1650 kHz	None	Top and bottom of T5 with a low-level signal generator input for maximum output.
RF and antenna coil	Signal generator to antenna input connector.	Tune for peak at 27.085 MHz	11 (27.085 MHz)	Top of T3 and T4 with a low-level signal generator input for maximum output.

Note: The T9 slug is adjusted for the best oscillator starting, using a minimum activity Channel 22 crystal. Top of T10 is adjusted for the best oscillator starting.

Table 7-1. Hallicrafters CB-21 receiver alignment.

Adjustment	Connection	Channel Crystal	Adjustment
Power output	Dummy load to antenna socket, power output indicator across load. 500-milliampere DC meter to TP1 (with jumper removed).	11 (27.085 MHz)	Tune T2, L5 and L6 for maximum output (below 5 watts) and final input (below 350 milliamperes at TP1).

Table 7-2. Hallicrafters CB-21 transmitter alignment.

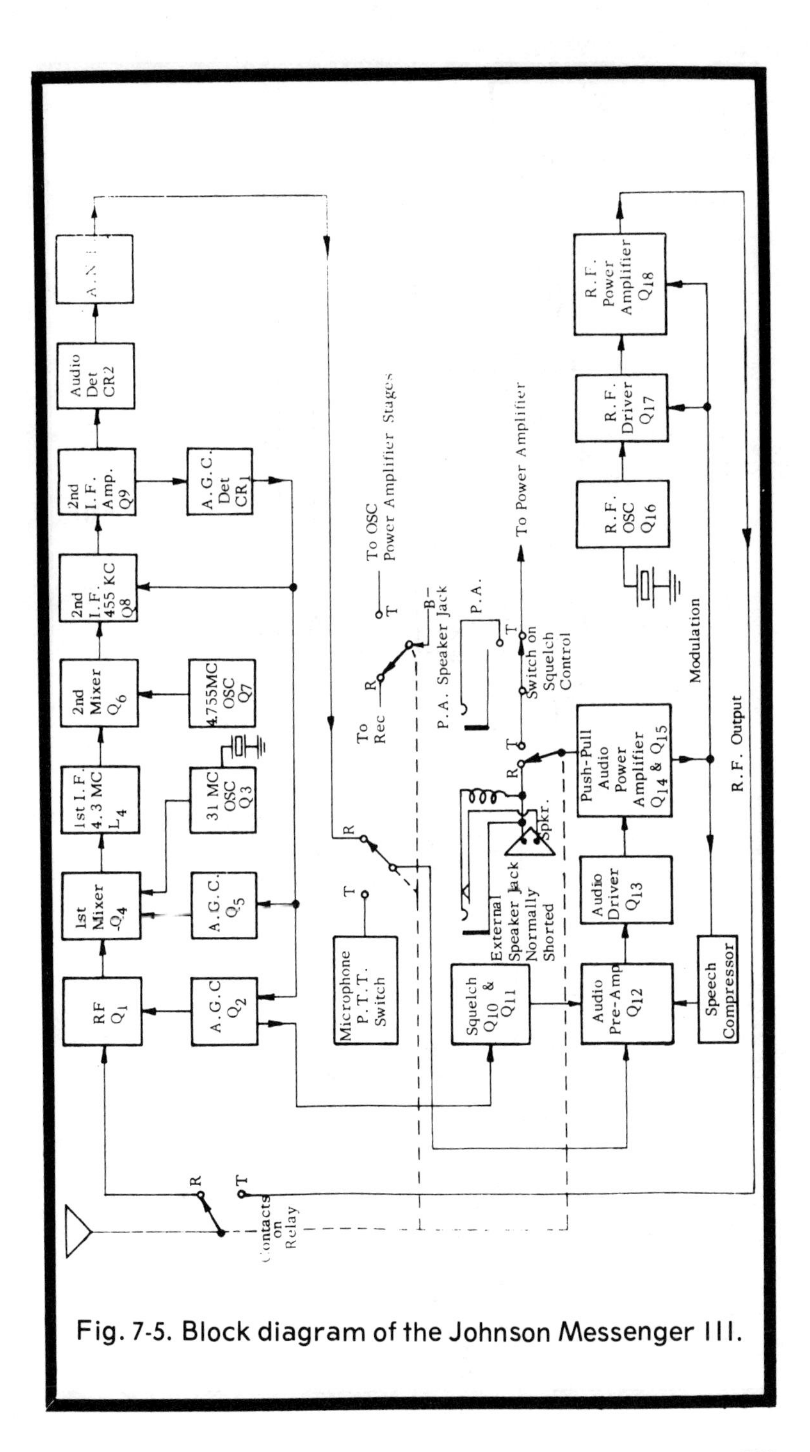

Fig. 7-5. Block diagram of the Johnson Messenger III.

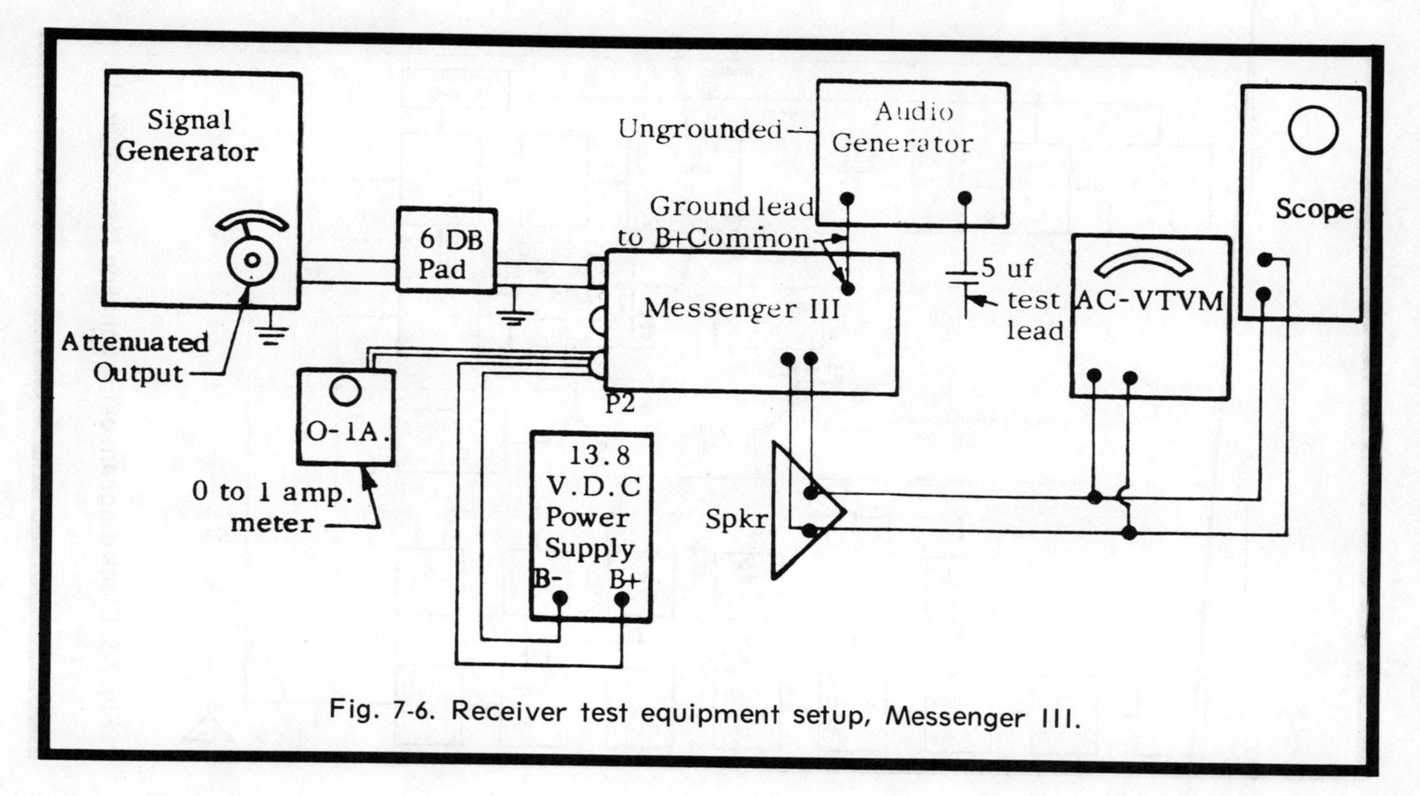

Fig. 7-6. Receiver test equipment setup, Messenger III.

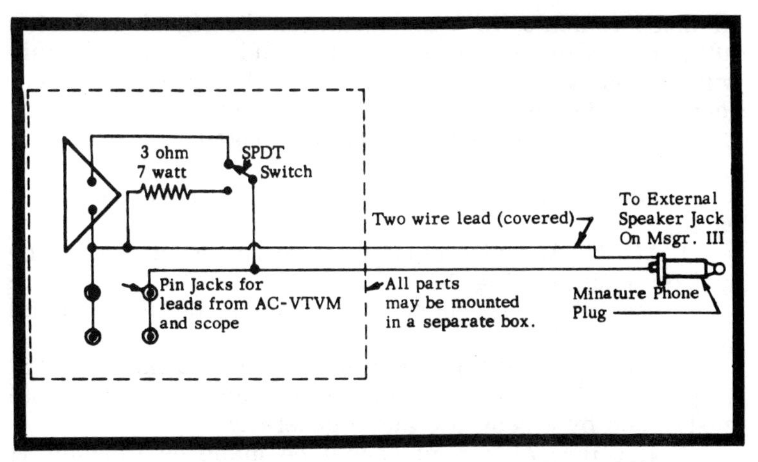

Fig. 7-7. AC VTVM and scope test setup, Messenger III.

The Messenger III transceiver is "floated" above ground and care must be exercised in the test equipment connection. The test equipment called for in the equipment list will work satisfactorily if connected as shown. The AC-VTVM and oscilloscope connections may be made as shown or a test cable may be used as shown in Fig. 7-7. The DC ammeter is connected to the power cable for transmitter readings as illustrated in Fig. 7-8, and the jumper between pins 5 and 6 must be replaced when the meter is not used. It may be necessary, in the use of the audio generator, to place a capacitor in series with the ground lead as well as the hot lead. The RF signal generator used for the receiver tests and alignment is connected to the receiver through a 6 db pad for

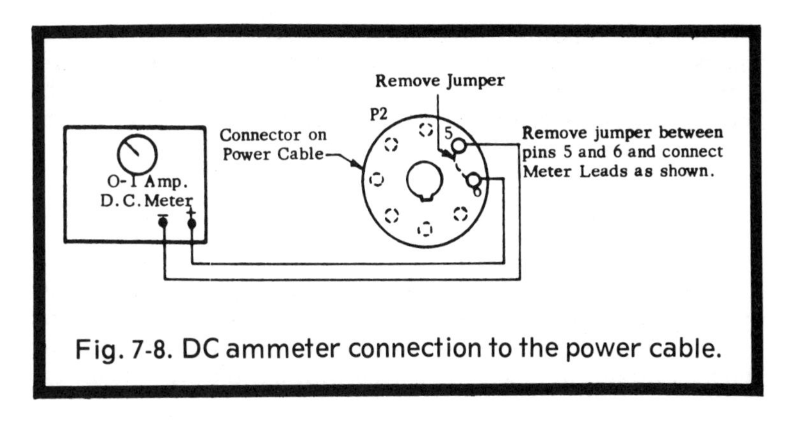

Fig. 7-8. DC ammeter connection to the power cable.

all measurements. In some instances the signal generator output may also be in series with a .1-mfd capacitor and this will be indicated where appropriate. The test equipment connections to the receiver for troubleshooting and alignment should be made as follows, except where noted in the text.

1. The DC supply is connected to J1 through the power cable.
2. AC VTVM and oscilloscope are connected across the speaker leads as in Fig. 7-6 or Fig. 7-7.
3. The 6 db pad is connected to the output of the RF signal generator.
4. The RF signal generator output is 1000-Hz modulated at 30 percent. This can be internal modulation or external modulation by a separate audio generator.
5. Receiver alignment should be made on Channel 11 (27.085 MHz).

C. Preliminary receiver checks

1. Connect the test equipment to the receiver as illustrated in Fig. 7-6.
2. Connect the signal generator to the antenna input.
3. Set the signal generator for a 1-microvolt output on Channel 11 (27.085 MHz).
4. Turn the volume control to maximum and squelch to minimum (CCW).
5. The current drain on the power supply for the receiver will be approximately 110 ma with a 13.8-volt DC power supply.
6. There should be at least +5 db of audio across the speaker.

If these conditions are not met, recommended receiver checks may be made as outlined in the following text. Note: The first check for trouble should be visual, then check the emitter currents of the stages by dividing the voltage at the emitter by the emitter resistance (Ie equals V3 over R3).

D. AGC section

The Messenger III utilizes a delayed, amplified AGC system. AGC is fed to the base of Q8 in the IF section and to the bases of Q5 and Q2, the AGC amplifiers. Q5 is in series with Q4, the first mixer, and Q2 is in series with Q1, the RF amplifier.

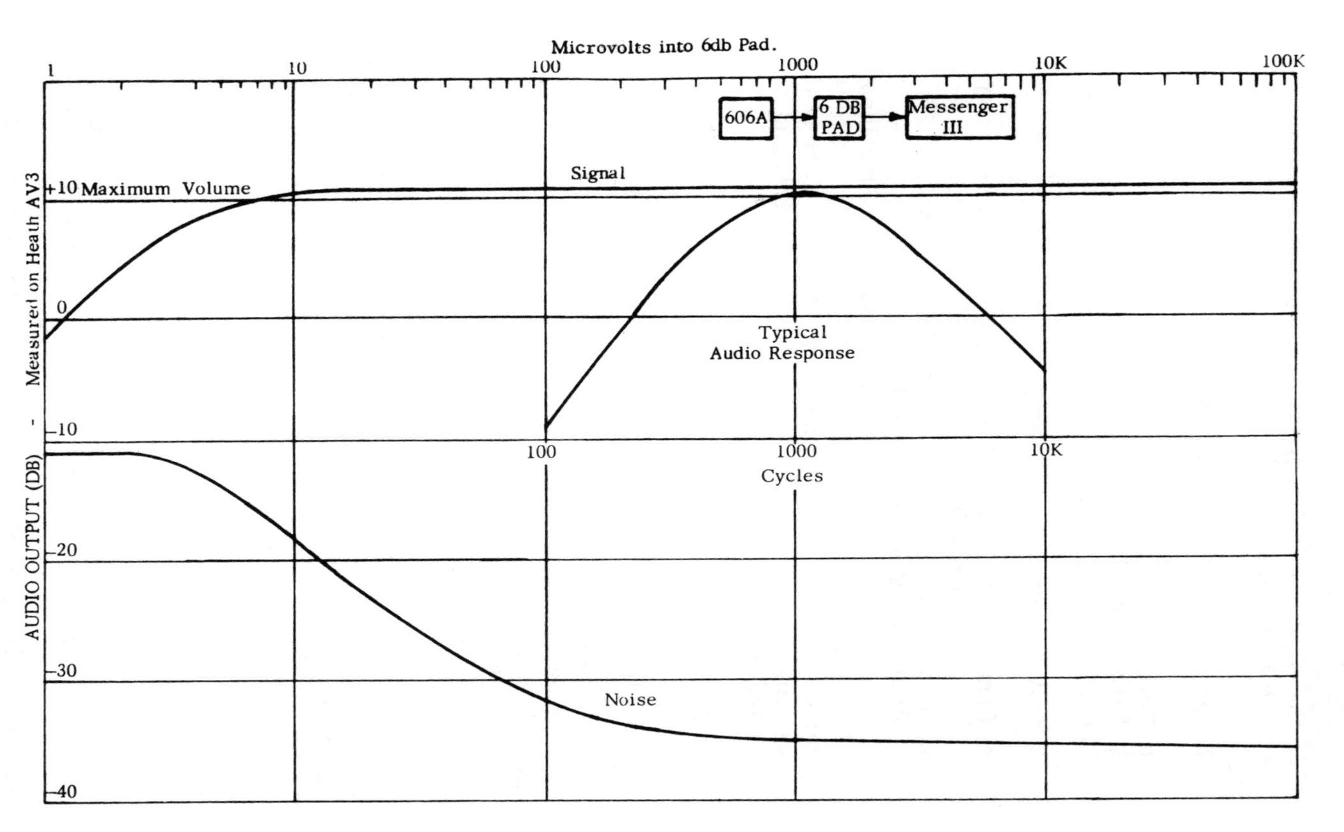

Fig. 7-9. Messenger III response curves.

Since the AGC may affect many stages, it is important to make checks on this stage first.

1. Absence of AGC will cause:

a. Severe overloading at high signal levels.
b. Erroneous voltage readings at the bases of Q1, Q2, Q4, Q5 and Q8.

2. AGC section troubleshooting

a. Connect the RF signal generator to the antenna terminal with a 1-microvolt output.
b. Connect a VTVM to the junction of R21 and R23.
c. Apply power to the receiver.
d. The AGC voltage will go more positive as the input signal is increased from 1 microvolt to .1 volt. See Fig. 7-9 for a typical AGC curve. (Audio output vs RF input.) If there is no voltage change at this point, check AGC detector CR1 and its associated network.

E. Squelch section troubleshooting

1. Connect a VTVM to the emitter of Q12. Apply power to the receiver.
2. While monitoring the VTVM, rotate the squelch control from minimum to maximum. The voltage should change from approximately -2.6 to -6.5 volts.
3. If the voltage does not change at Q12, check diode CR6 by bridging it with a new part.
4. An open CR6 will result in no squelch at all, and no change in voltage at the emitter of Q12.
5. If CR6 is not open, check the voltages at Q10 and Q11. This will aid in determining the presence of a defective transistor.
6. If diode CR6 is shorted, the voltage at Q11 will be normal and squelch will operate very slowly. The emitter of Q12 in this case would read very low at minimum squelch, and normal at maximum squelch. Check by replacing the diode.

F. Audio section troubleshooting

1. Connect the audio generator through a 5-mfd capacitor to the top of the volume control with an output of .0025 volts RMS, plus or minus 10 percent, at 1000 Hz (plus side to generator). The ground lead is connected to the common tie point (B+).

2. Set the audio control for +10 db on the AC VTVM and turn the squelch to minimum (CCW).

3. The output, as mentioned on the AC VTVM and scope, should be approximately +10 db undistorted. If this condition is not met, check the currents in stages Q12, Q13, Q14, and Q15. (Emitter current equals emitter voltage divided by emitter resistance.)

4. The voltages at the emitter and base of the push-pull transistors (Q14 and Q15) should be approximately the same. If one of the transistors shows no voltage difference between the emitter and base, it is probably defective. It would be best to remove the transistor and check it.

5. Parts replacement in audio stage. Replacement of Q14, Q15, R58, C51, C57, or CR5 will require the removal of L8 in order to gain access to these components. L8 is mounted on the side of the chassis and held in place by two screws.

Notes on the push-pull stage:

An open transistor in this stage is evidenced by severe audio distortion. A shorted transistor will cause R58 to burn and possibly blow a fuse. It may be possible to find the defective transistor by determining which has an excessively hot case. It is advisable to remove the transistors and check them with a good transistor tester and replace the defective part.

G. Second IF

1. Connect the test equipment as specified in Part B.
2. Troubleshooting

a. Apply power to the receiver, set the volume control for +10 db on the AC VTVM and turn the squelch to minimum.

b. Set the modulated RF signal generator to 455 kHz with an output of 200 microvolts.

c. Connect a .1-mfd capacitor in series with the output of the signal generator.

d. Connect the output of the signal generator to the base of Q6. There should be approximately +10 db of undistorted audio across the speaker terminals.

e. If the output is weak or distorted, check the IF stage voltages as well as the detector diode, CR2, and the noise limiter, CR3.

3. Alignment

a. Alignment is necessary only in the particular stage in which an IF transformer was replaced.

b. Inject the modulated 455-kHz signal as previously specified to the base of Q6. Ground the signal generator to the chassis rails.

c. Tune the second IF stage to the outside peak on both the top and bottom slugs as viewed from the top and bottom of the chassis respectively.

H. First IF section

1. Connect the test equipment as specified in Part B.
2. Troubleshooting

a. Set the modulated RF signal generator to 4.3 MHz with an output of 100 microvolts and ground the signal generator to the chassis rails.

b. Connect the signal generator to the base of Q4 through a .1-mfd capacitor.

c. The output across the speaker leads should be approximately +10 db of undistorted audio.

d. If these conditions are not met, check transformer L4 and the voltages at Q6 and Q7. Replace the defective components as necessary.

3. Alignment

If the replacement of L4, Q6 or Q7 was required, realign L4 as follows:

a. Inject a modulated 4.3-MHz signal to the base of Q4 through a .1-mfd capacitor as above.

b. Tune the bottom slug of L4 to the outside peak as viewed from the bottom of the receiver to obtain maximum output on the AC VTVM.

c. Tune the top slug to the outside peak as viewed from the top of the receiver to obtain maximum output on the AC VTVM.

I. RF amplifier and first mixer

1. Connect the test equipment as outlined in Part B.
2. Troubleshooting

a. Set the signal generator to 27.085 MHz (Channel 11) and connect it to the antenna input.

b. Set the modulated RF signal generator attenuator for a 1-microvolt output.

c. With maximum volume, the output read across the speaker terminals on the scope and AC VTVM should be at least +5 db undistorted audio.

d. If the receiver output is not normal, check the emitter of the oscillator, Q4, with an RF probe. There should be approximately .2 volts RF at this point.

e. If oscillator operation appears normal, check the voltages at Q1 and Q4 and replace if necessary.

f. Should it become necessary to replace either L1 or L2, proceed with the following alignment.

3. Alignment

a. If L2 is replaced, connect the RF signal generator to the antenna jack with a 1-microvolt output on Channel 11.

b. Peak L2 at the point where the slug is nearest the top of the coil form.

c. If L1 was replaced, turn the slug to the second peak as viewed from the top of the receiver.

d. Note the db reading on the AC VTVM after peaking L1 and detune this transformer (turn counterclockwise) for the

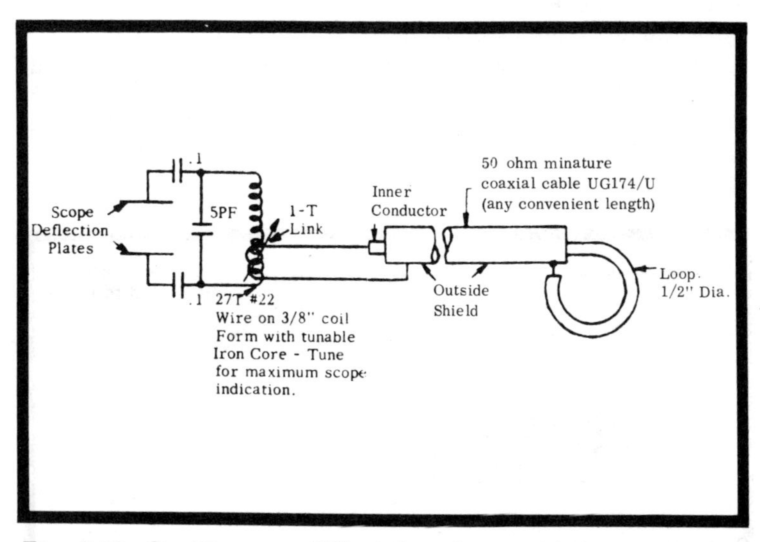

Fig. 7-10. Oscilloscope RF pickup loop and the method of connecting to the Messinger III.

best signal to noise relationship; this may be 1½ to 3 db removed from the peak gain.

Transmitter Troubleshooting

The following procedure is suggested as a guide for easy and quick trouble analysis of the Messenger III transmitter.

A. Equipment required

DC power supply, low transient regulated at 13.8 volts.

Oscilloscope, with RF pickup loop capable of direct connection to the vertical plates; see Fig. 7-10.

0-1 amp DC ammeter; connect to the power cable as shown in Fig. 7-8.

Dummy antenna; 50 ohms, 5 watts

Sine-wave generator; 1000 Hz, -40 db or higher.

B. Test equipment setup (Fig. 7-12)

1. The RF pickup loop is coupled to the desired stage to be checked.

2. Connect the 0-1 amp meter in series with pins 5 and 6 of power plug P2. These pins are jumpered in the plug as shown on the schematic. Be sure to replace this jumper when the meter is removed.

3. Connect the 50-ohm dummy antenna.

4. Connect the audio generator with .007 volts output through a .1-mfd blocking capacitor to pins 4 and 5 of plug P3. (Pins 4 and 5 are jumpered; connect to the jumper.) Connect the ground side of the generator to pin 6 of P3. See Fig. 7-11.

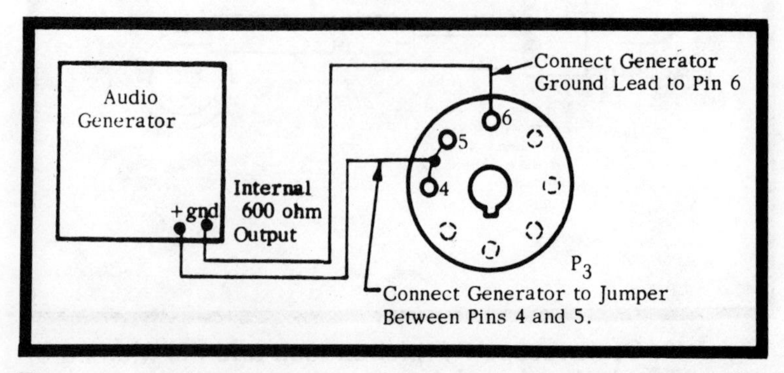

Fig. 7-11. Audio generator connection to P3, Messenger III.

5. Connect the DC power supply.

6. Turn the transceiver on and set the channel selector to Channel 11.

C. Preliminary transmitter check

1. Key the transmitter and check for an output of 3 watts minimum.

2. Feed in the 1000-Hz audio and monitor the output wave form on the scope. The output should be a clean sine wave. If the results of these checks show a defective transmitter, continue with the following procedure.

D. Initial transmitter adjustments

1. If the transmitter output is low, change the transmitter current with L16 and the peak power output with C79 to obtain the desired power point as shown in Fig. 7-14.

2. When the transmitter output is peaked to normal, modulate the carrier with .007 volts, 1000-Hz sine wave and check for "notching." (See Fig. 7-13.) If "notching" is present or the normal transmitter output cannot be obtained, proceed with the troubleshooting and alignment instructions.

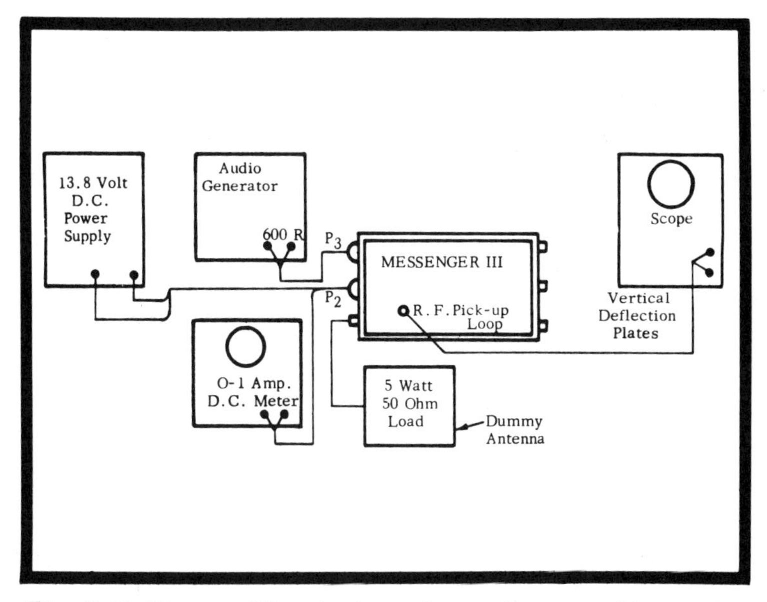

Fig. 7-12. Transmitter test equipment setup, Messenger III.

E. Oscillator stage.

1. Troubleshooting

a. Key the transmitter and check for the normal power output.

b. If the transmitter is inoperative, check the oscillator stage using the scope and RF pickup loop. A citizens band receiver may also be used to make a quick check of the oscillator. If no RF is present, check the voltages on Q16 and replace components as necessary.

2. Alignment

a. If replacement of transistor Q16 is required, it may be necessary to realign L11 to assure that the oscillator will start on all channels. Note: With no crystal in Channel B, switch capacity may couple the Channel A crystal to this position and cause oscillation. This is normal. A crystal installed in Channel B will operate normally.

b. Check for absence of "notching" while modulating the transmitter. (See Fig. 7-13.) If "notching" is present, refer to Section 4, Part H.

F. Driver stage

1. Troubleshooting

a. Key the transmitter; if the power output is low after adjusting C79 in part D, check the driver stage with the RF pickup loop.

b. If the RF is low at this point, check the voltages in this stage and replace defective components as necessary.

2. Alignment

a. If replacement of Q17 and L12 was necessary, realignment of this stage may be required.

b. Adjust L12 for maximum power output. This is a broad adjustment; tune for center of maximum.

c. Again check the transmitter output with a scope while modulating the carrier with a 1000-Hz sine wave to be sure there is no "notching." If "notching" is present, refer to Section 4, Part H.

G. Power amplifier stage (P.A.)

1. Troubleshooting

a. Key the transmitter; normal power output is approximately 3 watts minimum.

b. If C79 does not affect the output power and the driver stage is normal, check the voltages at Q18 and replace components as necessary.

c. Replacement of Q18, L15, or C79 may require realignment of the PA stage.

2. Alignment

a. If L15 was replaced, adjust the slug to ⅛ inch from being flush with the top of the coil form. Then proceed as follows. Note: **Do not touch L15** unless it has been replaced.

b. Key the transmitter and adjust C79 for maximum power output.

c. Check the PA current. If the current goes over 400 ma when C79 is peaked, retune L16 for less than 400 ma and readjust C79 for maximum power output. Repeat these adjustments for a minimum of 3 watts output at 400 ma or less.

H. "Notching" (due to mistuning) see Fig. 7-13.

1. Key the transmitter and modulate the carrier at 1000 Hz while monitoring the transmitter output on the scope as outlined in Part C.

2. While the transmitter is keyed, tune L11 until the "notching" disappears and the audio wave is symmetrical. If

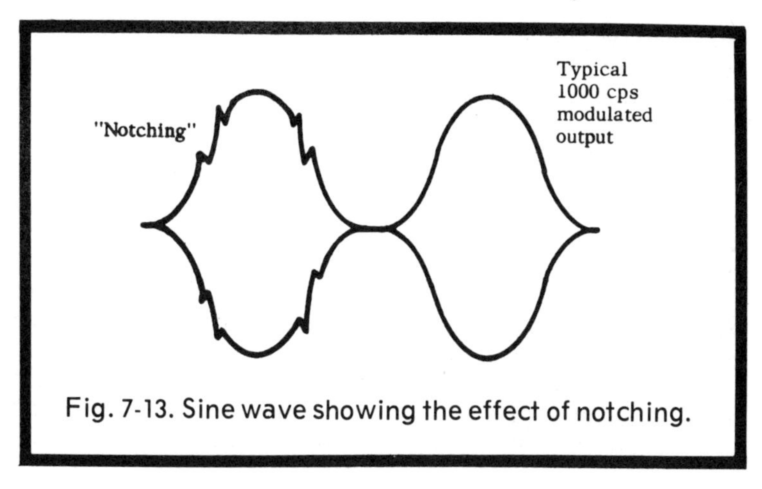

Fig. 7-13. Sine wave showing the effect of notching.

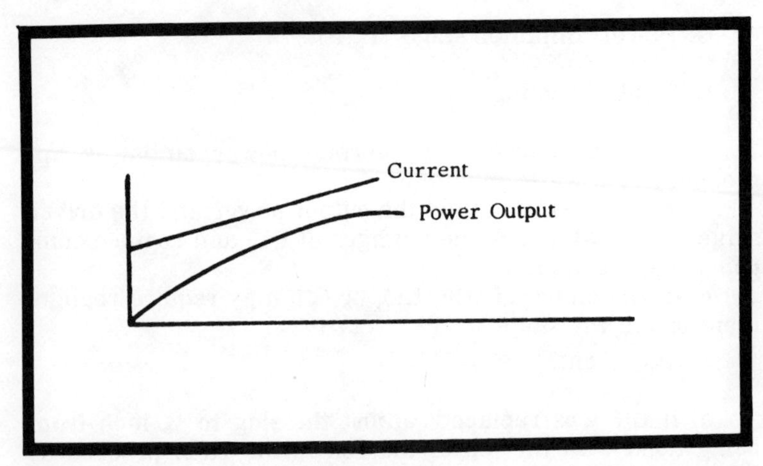

Fig. 7-14. Messenger III transmitter current-power curve.

L11 does not eliminate the "notching," it may also be necessary to make a slight adjustment of L12. Recheck to be sure that the oscillator is starting on all crystals as noted in Part E-2 above.

LAFAYETTE DYNA-COM 12A

The Dyna-Com 12A (foldout Panel E) is a full 5-watt hand-held transceiver designed for portable 12-channel 2-way radio communication. Housed in a rugged metal case (Fig. 7-15), the Dyna-Com 12A comprises a fully miniaturized transmitter and receiver, both crystal-controlled for precise, dependable operation. Fully transistorized, the unit employs a sensitive superheterodyne receiver circuit with one stage of RF and two stages of IF, and a 3-stage transmitter which uses rugged silicon transistors for longer life and more dependable performance.

A special feature in the transmitter is the full-time "range-boost" circuit which concentrates more audio power into the sidebands by providing high average modulation on all syllables. This results in a greater effective range of the transmitted signal. Other features include a mechanical filter for a sharp, selective reception, and adjustable squelch circuit which can be used to silence the receiver when no signals are being received, push-pull audio for high output and undistorted sound, AGC (automatic gain control) to prevent overloading on strong signals and to maintain uniform sound output, full-time automatic noise limiter, and an S-meter-RF

When ordering crystals, specify stock numbers of crystals (both stock numbers if both transmit and receive crystals are required) and channel on which operation is desired.

U.S. Channel	Transmitter Crystal Frequency MHz	Stock No.	Receiver Crystal Frequency MHz *	Stock No.
1	26.965	46-15019	26.510	46-10010
2	26.975	46-15027	26.520	46-10028
3	26.985	46-15035	26.530	46-10036
4	27.005	46-15043	26.550	46-10044
5	27.015	46-15050	26.560	46-10051
6	27.025	46-15068	26.570	46-10069
7	27.035	46-15076	26.580	46-10077
8	27.055	46-15084	26.600	46-10085
9	27.065	46-15092	26.610	46-10093
10	27.075	46-15100	26.620	46-10101
11	27.085	46-15118	26.630	46-10119
12	27.105	46-15126	26.650	46-10127
13	27.115	46-15134	26.660	46-10135
14	27.125	46-15142	26.670	46-10143
15	27.135	46-15159	26.680	46-10150
16	27.155	46-15167	26.700	46-10168
17	27.165	46-15175	26.710	46-10176
18	27.175	46-15183	26.720	46-10184
19	27.185	46-15191	26.730	46-10192
20	27.205	46-15209	26.750	46-10200
21	27.215	46-15217	26.760	46-10218
22	27.225	46-15225	26.770	46-10226
23	27.255	46-15233	26.800	46-10234

NOTE: Channel 23 is shared with Class "C" Radio Control.

* Although the receiver crystals are marked with a frequency which is 455 kHz (0.455 MHz) lower than the transmitter crystals, the receiver and transmitter both tune to channel frequency in column "Transmitter Crystal Frequency."

Table 7-3. Lafayette crystal chart.

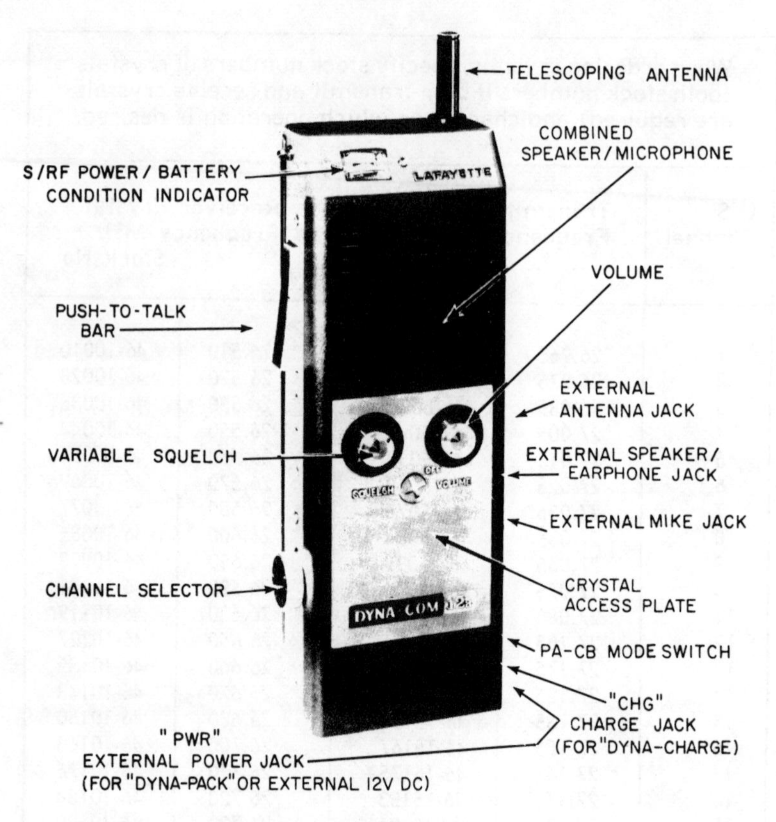

Fig. 7-15. Lafayette Dynacom 12A walkie-talkie.

power and battery condition indicator, plus a provision for PA (public address) operation.

Public Address Operation

Special provision has been made for public address (PA) operation, utilizing the microphone and audio stages in the transceiver. For PA operation, you would use an external 8-16 ohm speaker connected to the "SP-PH" jack. Set the PA-CB switch to PA, press the push-to-talk bar and talk into the front of the transceiver; your voice will be heard from the external speaker. Note: The volume control does not adjust the speaker output level in the PA mode.

Adding New Channels

The transceiver may be operated on any 12 channels (1 through 23) in the 27-MHz Citizens Band. Both receiver and transmitter are normally equipped with crystals for operation on Channel 10 unless otherwise ordered. Additional matched pairs of crystals may be ordered for operation on any of the other 11 channels.

Inserting new crystals

1. Loosen the screw in the lower front portion of the unit. Lift the cover plate off.

2. Insert the new matched pair of crystals into the sockets. Each crystal is marked with the channel (1-23), transmit or receive (T or R), and frequency. Make sure the transmitter (T) and receiver (R) crystals are installed in their correct sockets. If the crystals are reversed, the unit will transmit on one frequency and receive on another, both of which will be outside the Citizens Band, resulting in illegal operation.

3. Replace the cover plate and tighten the screw firmly. The unit now is ready for operation.

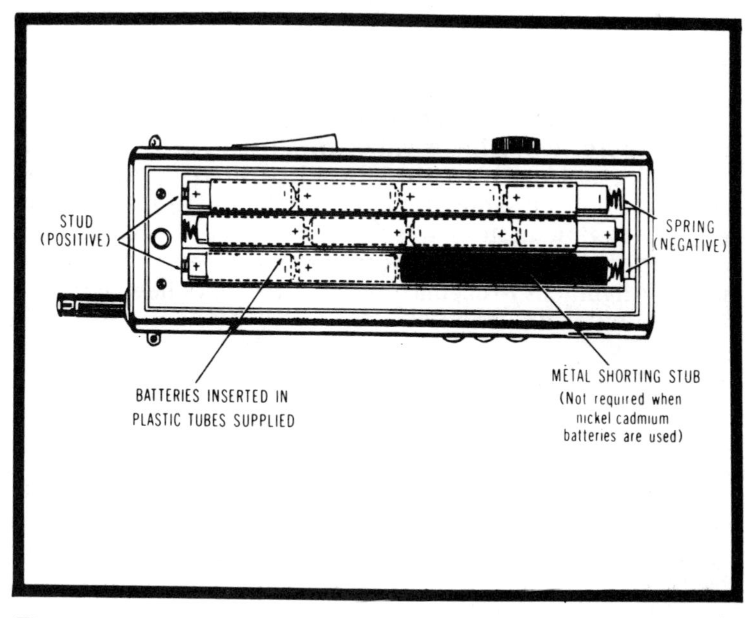

Fig. 7-16. Lafayette Dynacom 12A battery installation diagram.

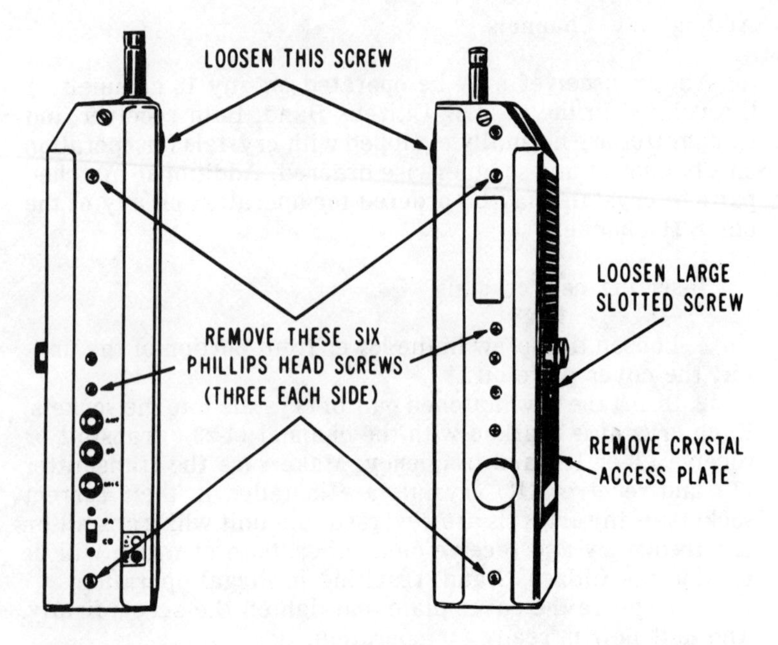

Fig. 7-17. Sides views of the Lafayette Dynacom 12A.

Service and Maintenance

The walkie-talkie incorporates a built-in 1½ amp fuse to provide protection in the event of an internal short circuit. If the walkie-talkie is completely inoperative (dead) even though the battery supply (or external DC supply) is known to be good, examination of this fuse should be made. To gain access to the fuse, remove the cover as indicated in the section under "Alignment Procedure." Fig. 7-16 shows how batteries are installed in the unit.

Simple Troubleshooting

If trouble is experienced with the unit, check as follows:

1. Make sure the squelch control is not advanced too far. If it is turned full on (fully clockwise), only very strong incoming signals can overcome the silencing action of the squelch circuit.

2. Make sure the PA-CB switch is in the CB position (for CB operation).

3. If trouble is experienced with transmission, make sure you are depressing the push-to-talk bar fully. Notice that only the lower portion of the bar depresses into the unit. Also notice whether or not the meter shows an indication of RF antenna power. If not, check the next two steps.

4. Make sure the channel selector is set to a position which is equipped with crystals.

5. Check the batteries for a weak or discharged condition. Replace or recharge the battery cells if necessary. Make sure the batteries are correctly inserted with regard to polarity.

6. Make sure the crystals are seated firmly. Check the frequencies or channel markings. Make sure that the transmit and receive crystals are in their proper positions.

Alignment

1. Remove the crystal access plate by loosening the large screw at the lower front of the unit (see Fig. 7-17).
2. Loosen the large screw at the back of the unit.
3. Remove the six small Phillips-head screws (three on each side of the unit).
4. Raise the main top cover carefully; there are two wire leads connected between the cover and the main unit (two speaker leads). Do not disconnect these leads.

To align the receiver portion of the transceiver, refer to foldout Panel E for the location of the alignment points and proceed as follows:

1. Connect an AC VTVM to the speaker terminals.
2. Connect a signal generator to the external antenna input jack and set the generator frequency to approximately the center of the CB band. This will vary with crystal selection, but Channel 10 will be suitable. Keep the generator output as low as possible at all times.
3. Turn the core of L10 clockwise until local oscillation stops. This will be indicated by a very low VTVM indication. Turn the core of L10 counterclockwise until the local oscillator begins to oscillate. This will be indicated by a sharp rise in the VTVM reading. Turn the core of L10 one-half turn further in the counterclockwise direction. This will ensure crystal oscillation under varying circuit conditions.

4. Adjust the cores of RF transformers L7, L8, L9 for a peak indication on the VTVM.

5. Adjust cores of MF, L11, L12 and L13 for the maximum VTVM indication.

6. Repeat Steps 4 and 5 until no further increase in the VTVM reading is achieved. The AC VTVM reading should be approximately 2.5v for 1 microvolt (30 percent modulated) signal generator input.

To calibrate the battery meter, apply 15 volts DC to the transceiver's power input connections and adjust potentiometer VR-5 for a full-scale reading in the extreme right edge of the center on the built-in battery condition meter.

To perform transmitter alignment

1. Connect a 50-ohm wattmeter to the external antenna jack on the transceiver.

2. Set the transceiver's channel selector to any position which contains a crystal socket with the proper transmitting crystal; if possible, use a channel approximately in the center of the CB band.

3. Depress the microphone button and adjust the core of L1 clockwise until no power output is indicated on the wattmeter (stop at the precise point). Turn the core of L1 counterclockwise until oscillation starts (wattmeter indicates RF output with this setting). Turn counterclockwise one more turn.

4. Adjust the core of L2 for the maximum RF output as indicated on the wattmeter.

5. Expand or compress the turns of L3 and L4 for the maximum power output as indicated on the wattmeter. Note: L5 is factory adjusted and should not be adjusted.

6. To tune the whip antenna, adjust L6 for maximum output as indicated on the built-in output meter. Be sure that the whip is fully extended and that the wattmeter is disconnected from the external antenna jack.

LAFAYETTE MODEL HB-23A

The following specifications apply to the HB-23A:

Receiver circuit type: Dual conversion superheterodyne with an RF stage and a 455-kHz mechanical filter.

Frequency: Up to 23 crystal-controlled channels in the 27-MHz Citizens Band (normally shipped with crystals for

Channel Frequency (MHz)	FOR TRANSMIT Crystals Used (MHz)	Socket	FOR RECEIVE Crystals Used (MHz)
1 26.965	Xtal 1: 38.275 Xtal 14: 11.310	A *	Xtal 1: 38.275 Xtal 16: 11.765
2 26.975	Xtal 2: 38.285 Xtal 14: 11.310	B *	Xtal 2: 38.285 Xtal 16: 11.765
3 26.985	Xtal 3: 38.295 Xtal 14: 11.310	C *	Xtal 3: 38.295 Xtal 16: 11.765
4 27.005	Xtal 4: 38.315 Xtal 14: 11.310	D *	Xtal 4: 38.315 Xtal 16: 11.765
5 27.015	Xtal 1: 38.275 Xtal 13: 11.260	A *	Xtal 1: 38.275 Xtal 15: 11.715
6 27.025	Xtal 2: 38.285 Xtal 13: 11.260	B *	Xtal 2: 38.285 Xtal 15: 11.715
7 27.035	Xtal 3: 38.295 Xtal 13: 11.260	C *	Xtal 3: 38.295 Xtal 15: 11.715
8 27.055	Xtal 4: 38.315 Xtal 13: 11.260	D *	Xtal 4: 38.315 Xtal 15: 11.715
9 27.065	Xtal 5: 38.375 Xtal 14: 11.310	E ** *	Xtal 5: 38.375 Xtal 16: 11.765
10 27.075	Xtal 6: 38.385 Xtal 14: 11.310	F *	Xtal 6: 38.385 Xtal 16: 11.765
11 27.085	Xtal 7: 38.395 Xtal 14: 11.310	G *	Xtal 7: 38.395 Xtal 16: 11.765
12 27.105	Xtal 8: 38.415 Xtal 14: 11.310	H *	Xtal 8: 38.415 Xtal 16: 11.765
13 27.115	Xtal 5: 38.375 Xtal 13: 11.260	E ** *	Xtal 5: 38.375 Xtal 15: 11.715
14 27.125	Xtal 6: 38.385 Xtal 13: 11.260	F *	Xtal 6: 38.385 Xtal 15: 11.715
15 27.135	Xtal 7: 38.395 Xtal 13: 11.260	G *	Xtal 7: 38.395 Xtal 15: 11.715
16 27.155	Xtal 8: 38.415 Xtal 13: 11.260	H *	Xtal 8: 38.415 Xtal 15: 11.715
17 27.165	Xtal 9: 38.475 Xtal 14: 11.310	I *	Xtal 9: 38.475 Xtal 16: 11.765
18 27.175	Xtal 10: 38.485 Xtal 14: 11.310	J *	Xtal 10: 38.485 Xtal 16: 11.765
19 27.185	Xtal 11: 38.495 Xtal 14: 11.310	K= *	Xtal 11: 38.495 Xtal 16: 11.765
20 27.205	Xtal 12: 38.515 Xtal 14: 11.310	L *	Xtal 12: 38.515 Xtal 16: 11.765
21 27.215	Xtal 9: 38.475 Xtal 13: 11.260	I *	Xtal 9: 38.475 Xtal 15: 11.715
22 27.225	Xtal 10: 38.485 Xtal 13: 11.260	J *	Xtal 10: 38.485 Xtal 15: 11.715
23 27.255	Xtal 12: 38.515 Xtal 13: 11.260	L *	Xtal 12: 38.515 Xtal 16: 11.765

* Permanently wired into the circuit—not plug-in type.
= Crystal in socket K is factory-inserted and is not removable.
** Crystal in socket E is factory-inserted, but removable.

Table 7-4. Crystals used for transmit and receive functions for each channel.

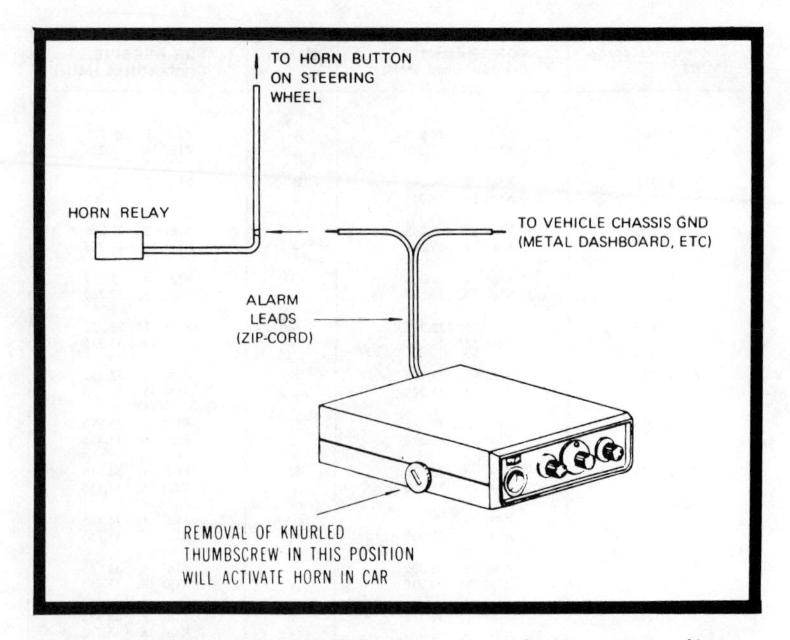

Fig. 7-18. Lafayette HB-23A burglar alarm connections.

operation on Channels 9, 13 and 19).

Sensitivity: 0.7 microvolt for 10 db S + N-N ratio.

Selectivity: 6 db down at 6 kHz, 45 db down at plus or minus 8 kHz.

Intermediate frequency: 1st IF: 11.260 or 11.310 MHz; 2nd IF: 455 kHz.

Audio output: 3 watts maximum into external speaker jack.

Receiving current drain: 100 milliamps on standby (no signal).

Transmitter frequency: Up to 23 crystal-controlled channels in the 27-MHz Citizens Band (normally shipped with crystals for operation on Channels 9, 13 and 19).

Power input: 5 watts.

Emission: 8A3.

Spurious response rejection: All harmonic and spurious suppression greater than FCC and D.O.T. requirements.

Modulation: AM, 90 percent typical.

Range boost: Yields high average modulation at average voice levels.

Transmitting current drain: Less than 1 amp.

Antenna: Nominal 50 ohms impedance (may be used with 30-100 ohm antennas).

Even though these units are equipped with crystals for Channels 9, 13 and 19, ten additional crystals may be inserted into the synthesizer circuit for operation on the remaining 20 channels in the Citizens Band (each additional crystal provides two transmit-receive channels). Designed and built for reliable, trouble-free performance, the HB-23A uses rugged, heat-resistant transistors in all critical areas. Current drain on 12 volts DC is exceptionally low, permitting continuous mobile operation for long periods of time, even with the automobile's motor switched off.

The BH-23A is designed to operate from 11.5 to 14.5 volts DC in a positive or negative ground electrical system, but it may also be operated from 105-120 volts, 50-60 Hz AC when used with the optional solid-state AC power supply unit Model HB-502A. The HB-23A may also be operated as a full 5-watt portable 2-way radio when used with optional portable battery pack Model HB-507. This pack, which operates from eight 1½ volt "C" size dry cells or ten "C" size nickel cadmium cells, is equipped with telescoping whip antenna, battery condition meter, microphone clip and shoulder strap.

Burglar Alarm Circuit

The transceiver is equipped with a built-in switching circuit designed to discourage the removal of the unit by unauthorized persons. This switching circuit which operates upon removal of the left-hand knurled thumbscrew from the transceiver, terminates in two zip-cord leads at the rear of the unit. When these leads are connected to the automobile horn circuit and ground, as indicated in Fig. 7-18, any attempt to remove the thumbscrew in order to remove the transceiver from its mounting will cause the automobile horn to blow.

Connecting the Alarm Leads

Note: Be sure to mount the transceiver in the automobile with the knurled securing screws at each side tightened before connecting the alarm leads to the horn circuit. Connect one of the two zip-cord leads from the rear of the transceiver to the chassis of the vehicle (or metal dashboard, etc.). Connect the other lead to the same terminal on the horn relay to which the horn button lead from the steering wheel is connected. If you are in doubt about this connection, check with your local automotive service station or garage. Be sure to install these leads as inconspicuously as possible so that they are concealed from view.

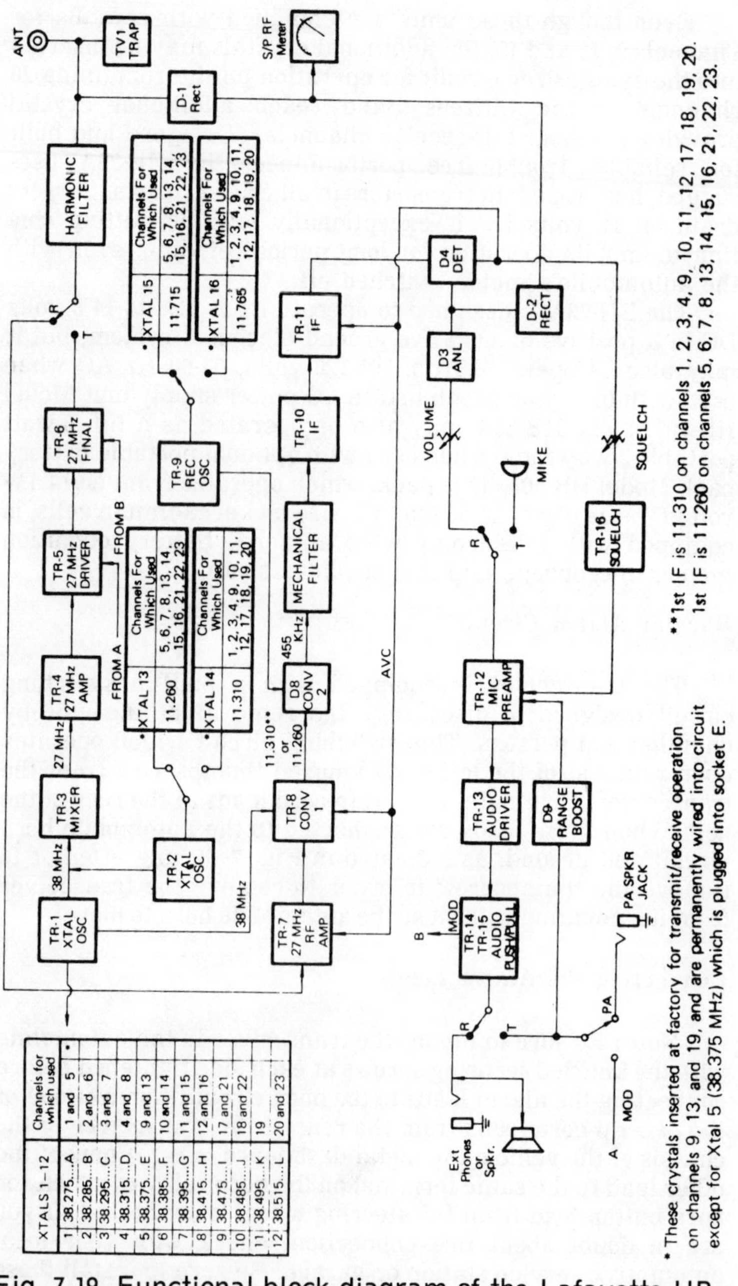

XTALS 1–12		Channels for which used
1	38.275....A	1 and 5
2	38.285....B	2 and 6
3	38.295....C	3 and 7
4	38.315....D	4 and 8
5	38.375....E	9 and 13
6	38.385....F	10 and 14
7	38.395....G	11 and 15
8	38.415....H	12 and 16
9	38.475....I	17 and 21
10	38.485....J	18 and 22
11	38.495....K	19
12	38.515....L	20 and 23

	Channels For Which Used
•XTAL 13 11.260	5, 6, 7, 8, 13, 14, 15, 16, 21, 22, 23
•XTAL 14 11.310	1, 2, 3, 4, 9, 10, 11, 12, 17, 18, 19, 20

	Channels For Which Used
•XTAL 15 11.715	5, 6, 7, 8, 13, 14, 15, 16, 21, 22, 23
•XTAL 16 11.765	1, 2, 3, 4, 9, 10, 11, 12, 17, 18, 19, 20

Fig. 7-19. Functional block diagram of the Lafayette HB-23A transceiver.

Crystal Synthesizing System

This transceiver incorporates a frequency synthesizing system which employs 16 crystals to produce 23 transmitting channels and 23 receiving channels. This represents a great reduction in the number of crystals required when compared to systems which employ one transmit crystal and one receive crystal for each channel (this system would require 46 crystals for 23 channel coverage). The system used in the HB-23A is unique in its operation, even when compared to other frequency synthesizing circuits. First, 10 of the crystals are optional plug-in types and not permanently wired into the circuit (11 crystals are actually plug-in types, but one crystal is factory-installed). Second, the circuit has been so designed as to permit the future addition of these 10 crystals (one at a time, if desired) to activate additional channels. This makes it possible to supply the transceiver working on three basic channels (9, 13 and 19), allowing the owner to activate additional channels at any future time, as desired. This arrangement offers the advantage of reduced initial cost without sacrificing the main advantage that frequency synthesis provides; namely, low total crystal cost for multiple channels. Each of the 10 optional crystals activates two transmit and two receive channels, as shown in Table 7-5. With the insertion of all 10 crystals, 20 additional channels are added to the three basic operating channels to provide full 23-channel transmit-receive operation.

In order to understand the manner in which the system operates, it is first necessary to know which crystals are selected for each channel. This information is provided in Table 7-4 and in the functional block diagram of the entire transceiver in Fig. 7-19. Figure 7-20 is a partial block diagram showing the various frequency converions that take place during transmit and receive on Channel 1 (26.965 MHz). Table 7-4 shows that the crystals used in the transmit mode are Xtal 1 (38.275) and Xtal 14 (11.310), and those used in the receive mode are Xtal 1 (38.275) and Xtal 16 (11.765).

In the transmit mode, TR-1 is connected to Xtal 1 and TR-2 is connected to Xtal 14. The output frequencies of these two crystal oscillators (38.275 and 11.310 MHz, respectively) are fed to the mixer stage TR-3. The output of the mixer stage contains tuned circuits which will pass only frequencies in the 27-MHz range. Thus, only the difference frequency (38.275 - 11.310 equals 26.965 MHz) is applied to TR-4, etc. This system of "beating" of two crystal frequencies takes place on every channel, the appropriate pair of crystals (for TR-1 and TR-2)

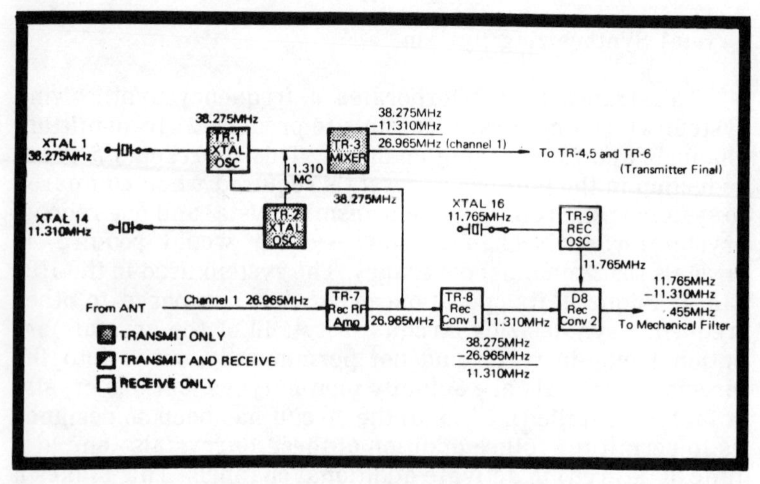

Fig. 7-20. Operation of the frequency synthesizing system in transmit and receive modes on Channel 1. Notice that Xtal 1 and Xtal 14 are used to produce the channel frequency during transmit. During receive, the Xtal 1 frequency is mixed with the incoming signal to produce the first IF of 11.310 MHz; the frequency of Xtal 16 is mixed with 11.310 MHz at D8 to produce the second IF of 455 kHz.

being automatically selected in each case to produce a difference frequency which is the channel frequency. In the receive mode, the output of TR-1 (38.275) is mixed with the incoming signal of 26.965 (which has been passed through RF amplifier TR-7).

The output of converter TR-8 contains tuned circuits which will pass only frequencies in the 11-MHz range. Thus, only the difference frequency (38.275 - 26.965 equals 11.310 MHz) is applied to the following stage, D8. TR-9, which has been automatically connected to Xtal 16 when Channel 1 was selected, produces an output frequency of 11.765 MHz which is also fed to D8. The output of D8 contains a sharply tuned mechanical filter which will pass only 455 kHz, this being the difference of the two frequencies (11.765 - 11.310 equals 455 kHz). This system of converting the incoming frequency twice to produce a second IF of 455 kHz takes place on every channel, the appropriate pair of crystals (for TR-1 and TR-9) being automatically selected in each case. It should be noted that because of the particular frequencies chosen for Xtals 1-12, the difference frequency at the output of TR-8, or first IF,

Xtal NO.	Socket	FREQUENCY	CHANNELS FOR WHICH USED	
			TRANSMIT	RECEIVE
1	A	38.275	1 and 5	1 and 5
2	B	38.285	2 and 6	2 and 6
3	C	38.295	3 and 7	3 and 7
4	D	38.315	4 and 8	4 and 8
5 *	E	38.375	9 and 13	9 and 13
6	F	38.385	10 and 14	10 and 14
7	G	38.395	11 and 15	11 and 15
8	H	38.415	12 and 16	12 and 16
9	I	38.475	17 and 21	17 and 21
10	J	38.485	18 and 22	18 and 22
11*	K	38.495	19	19
12	L	38.515	20 and 23	20 and 23
13*		11.260	5, 6, 7, 8, 13, 14, 15, 16, 21, 22, 23.	----------
14*		11.310	1, 2, 3, 4, 9, 10, 11, 12, 17, 18, 19, 20.	----------
15*		11.715	----------	5, 6, 7, 8, 13, 14, 15, 16, 21, 22, 23.
16*		11.765	----------	1, 2, 3, 4, 9, 10, 11, 12, 17, 18, 19, 20.

* Transceiver is factory-equipped with these crystals; all others are plug-in types available as optional extras.

Table 7-5. Cross-reference of crystal functions.

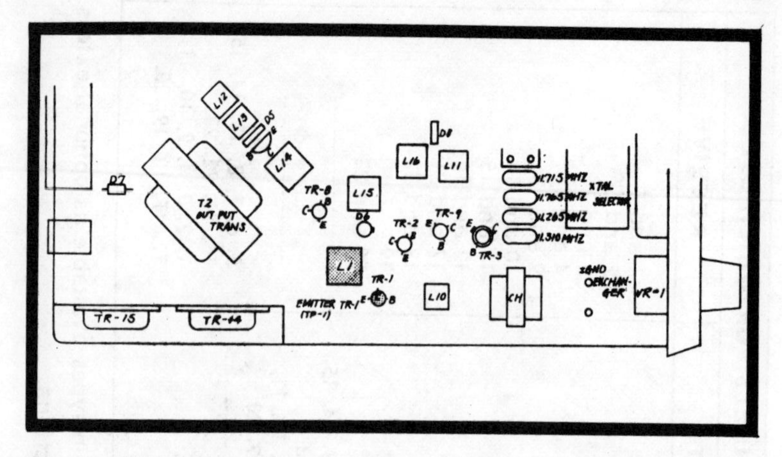

Fig. 7-21. Partial layout diagram of the Lafayette HB-23A chassis with emphasis on the location of L1.

will be either 11.310 (as in the case of Channel 1) or 11.260 MHz, depending on the channel in use.

Table 7-5 shows the operating mode and number of channels for which each of the 16 crystals is used (this includes the 10 optional plug-in crystals). This information is identical to that found in Table 7-4 except that it is presented in a form suitable for pin-pointing the failure of a crystal. For example, failure of crystal 14 will cause the transmitter to be inoperative on Channels 1, 2, 3, 4, 9, 10, 11, 12, 17, 18, 19 and 20, although the receiver will function normally on these same channels. When malfunctions occur on a selective number of channels in this manner, failure of a crystal or its associated wiring should be suspected initially.

Alignment

The transceiver has been fully aligned at the factory and does not normally require further adjustment. When necessary, however, the receiver and transmitter may be aligned as indicated.

1. Remove the mounting brackets, etc.

2. Turn the unit upside-down with the speaker grille facing upward. Remove the two upper screws on each side of the unit. Detach the speaker-mounted cover carefully because the speaker is connected to the main chassis by means of two leads terminated with push-on lugs.

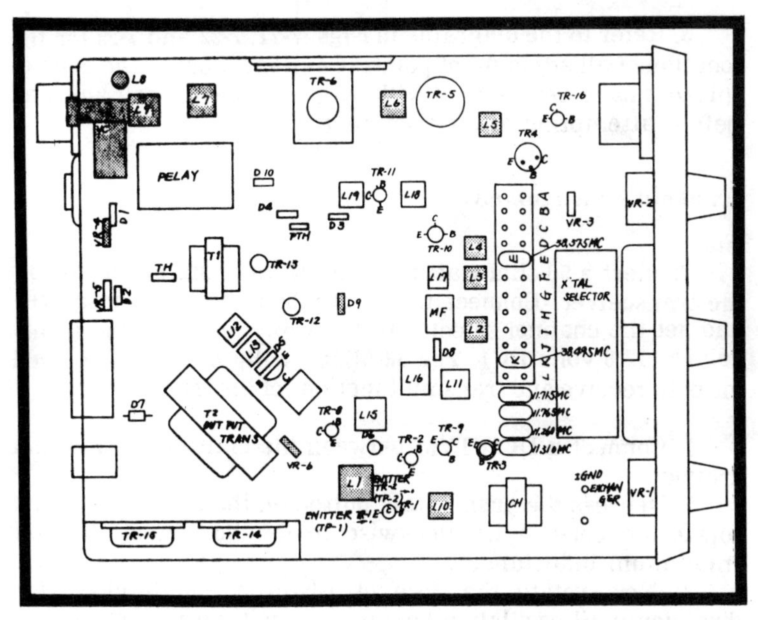

Fig. 7-22. Lafayette HB-23A layout with transmitter adjustments shaded.

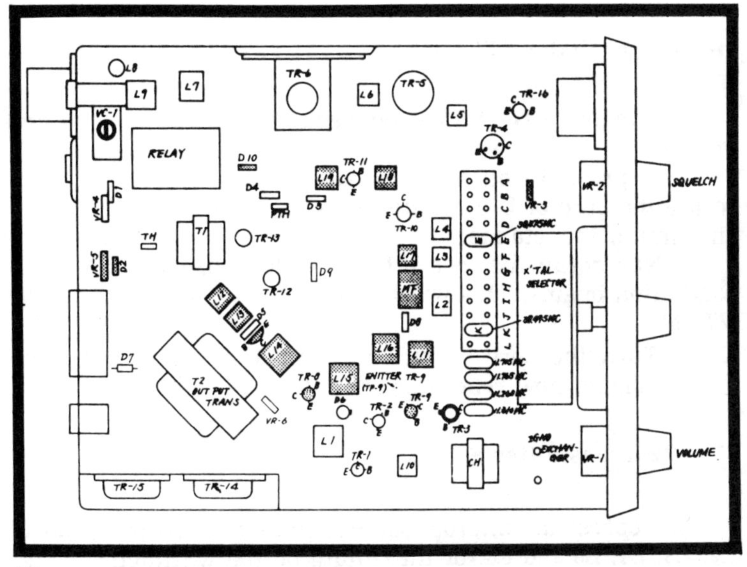

Fig. 7-23. BH-23A chassis layout with receiver adjustments shaded.

3. Refer to the diagrams in Figs. 7-21, 7-22 and 7-23 for the location of all adjustment points. Caution: Most of the coils in this unit have been sealed with wax. Be sure to melt any wax before attempting any adjustment.

Transmitter Alignment

Connect a 50-ohm wattmeter to the antenna connector on the transceiver. Connect the microphone to the transceiver and set the channel selector to 13. Apply power to the transceiver (12.6 volts DC). The 38-MHz oscillator (TR-1) is used for both receive and transmit functions of the transceiver.

1. Connect a DC VTVM between the emitter of TR-1 and ground.
2. Depress the push-to-talk button on the microphone and rotate the core of L1 clockwise to the bottom of the coil (maximum inductance).
3. Now rotate the core of L1 in a counterclockwise direction until oscillation begins. This will be indicated by a reading on the VTVM.
4. Turn the core ½ turn more in a counterclockwise direction. The VTVM should now read approximately 2.3 volts.

11-MHz Oscillator (TR-2)

1. Connect a DC VTVM between the emitter of TR-2 and ground.
2. Depress the push-to-talk button on the microphone and rotate the core of L10 counterclockwise to the top of the coil (minimum inductance).
3. Now rotate the core of L10 in a clockwise direction until oscillation begins. This will be indicated by a reading on the VTVM.
4. Turn the core ½ turn in a clockwise direction. The VTVM should now read approximately 2.5 volts.

27-MHz Transmitter Stages

1. Depress the microphone button and adjust the cores of L2, L3, L4, L5 and L6 for maximum on the wattmeter. Note: The adjustment of L5 is fairly critical. Misadjustment of this coil can reduce the transmitter output to zero.

2. Check the power output on Channels 9, 13 and 19. If low on some channels, readjust L2, L3, L4, L5 and L6 for equal output on Channels 9 and 19.

3. Adjust L7, L8 and VC for maximum output on the wattmeter. Note: L7 is adjusted by either compressing or expanding the coil turns. Use a nonmetallic tuning tool to spread the wire turns.

4. Adjust VR-4 so that the transceiver meter reads the same as on the wattmeter.

5. The transceiver may be repeaked for maximum RF power output at the actual installation with the antenna connected by readjusting VC-1 for maximum radiated power on an RF field strength meter.

TVI Adjustment

1. Use a TV receiver set to Channel 2 as an indicator.

2. Depress the transceiver microphone button and adjust L9 (rear of transceiver) for minimum interference on the TV receiver.

455-kHz IF

1. Connect the signal generator to the cathode of D8.

2. Connect the AC VTVM to the speaker terminals.

3. Set the signal generator to 455 kHz, plus or minus 0.5 kHz.

4. Apply power to the unit and adjust the signal generator output to produce a reading of 0.5 volts on the AC VTVM.

5. Adjust the mechanical filter (MF), L17, L18 and L19 for maximum output on the VTVM. Note: Reduce the output of the signal generator as necessary to keep the VTVM reading around 0.5 volts.

Receive Oscillator (TR-9)

1. Connect a DC VTVM between the emitter of TR-9 and ground.

2. Rotate the core of L11 counterclockwise to the top of the coil (minimum inductance).

3. Now turn the core of L11 in a clockwise direction until the oscillation begins. This will be indicated by a reading on the VTVM.

4. Turn the core ½ turn more in a clockwise direction. The VTVM should now read approximately 1.5 volts.

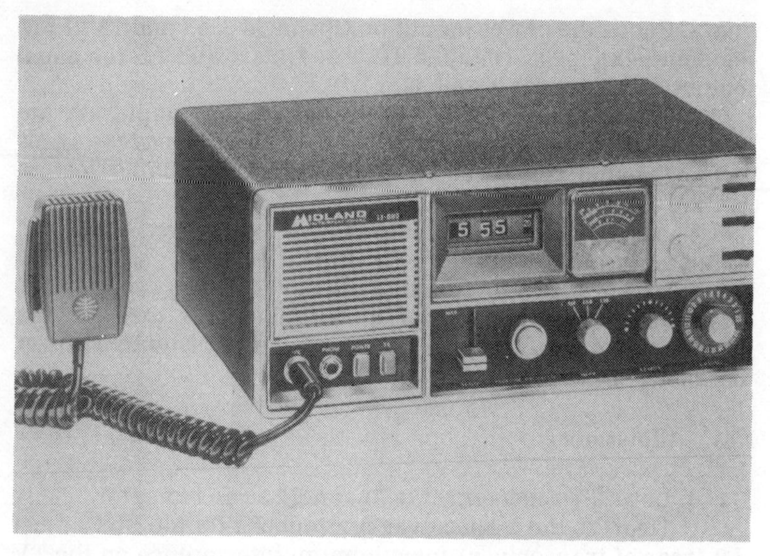

Fig. 7-24. Midland Model 13-880.

11.260 MHz and 11.310 MHz IF

1. Connect the signal generator to the base of transistor TR-8.
2. Connect the AC VTVM to the speaker terminals.
3. Set the signal generator to 11.285 MHz.
4. Adjust L15 and L16 for maximum output as read on the VTVM. Reduce the signal generator output as necessary to keep the VTVM reading around 0.5 volts.

RF Alignment

1. Connect the signal generator to the antenna connector.
2. Connect the AC VTVM across the speaker terminals.
3. Set the signal generator to 27.115 MHz, modulated 30 percent with a 1-kHz tone. Set the signal generator output to 10 microvolts.
4. Set the transceiver to Channel 13 and vary the signal generator frequency around 27.115 MHz to produce a maximum reading on the AC VTVM. Leave the generator at this point.
5. Adjust L12, L13 and L14 to produce maximum output on the AC VTVM.
6. Reduce the generator output to approximately 1 microvolt. Adjust L12, L13, L14, L15, L16, MF (mechanical

filter), L17, L18 and L19 for a maximum reading on the VTVM. Repeat until no further improvement is noted.

7. Increase the generator output to 100 microvolts and adjust VR-5 so that the transceiver meter reads S-9.

8. Squelch alignment: Receive a signal with a signal strength of 80 db from the signal generator. Turn the squelch control full clockwise, then adjust VR-3 until the signal is heard.

MIDLAND 13-880

The Midland Model 13-880 is designed to be used as either a base station or a mobile unit. It can be powered from either 117v AC or 12v DC. The unit has three modes of operation: upper single sideband, lower single sideband and high-level Class B AM. It is equipped with a digital clock and a front panel meter that reads power output, SWR and incoming signal level (Fig. 7-24). The SWR position also has an external calibration control to give a more accurate reading under varying conditions.

The Model 13-880 uses a synthesizer with a total of 14 crystals to obtain all 23 channels. Two additional crystals are used for IF conversion in the SSB mode of operation.

The transmitter consists of an integrated circuit mixer, a buffer, a driver and a final RF amplifier. The three stages are tuned in the AM mode for maximum output into a 50-ohm dummy load. The final stage must then be checked by inserting a milliammeter in the circuit across TPI and TPZ. This reading should be 10 milliamps with no modulation present. To properly match the final stage to an antenna, coils L2, L3, and L4 are adjusted. The only tuning required in the SSB mode is the carrier balance circuit. With no modulation present, controls VR2 and CV1 should be adjusted for a minimum of RF output. There should be less than 0.2 volts RMS present at the antenna terminal when these controls are properly adjusted.

The receiver uses double conversion with the IF frequencies at 7.8 MHz and 455 kHz. With the exception of the second IF stages, all tuning should be done with a very accurate signal generator and, if possible, a frequency counter. Proper operation of the receiver in the SSB mode depends upon the accuracy of the high IF stages and the crystal oscillator supplying the mixing frequencies.

To troubleshoot the receiver section, a stage gain check may be made by feeding a known value of signal into each stage and checking for the required output at the collector of Q12, the first audio amplifier, with a VTVM. A chart supplied with this unit gives input signal levels and output levels for all of the amplifier stages in the receiver.

The synthesizer frequency adjustments must be made using the chart in the instruction manual that gives the various combinations required to obtain the 23 frequencies required. During this procedure the 7-MHz oscillator must be disabled by grounding the base of the upper side-band oscillator, Q203. By adjusting trimmer CV10 through CV15 on Channels 1, 5, 9, 13, 17 and 21 all of the remaining channels will be on frequency. The upper and lower sideband oscillators must also be adjusted separately to ensure accurate reception of the sideband signals.

PERSONAL MESSENGER

This unit is typical of the smaller hand-held portable transceivers used in the Citizens Band. The input to the final stage is 1½ watts with an output of approximately 0.7 watts. (See the schematic on foldout Panel A.) This is a single-channel unit. The transmitter employs only two transistors to generate the RF power. The crystal oscillator (Q10) generates the RF signal which is coupled through the tuned circuit (C33, L7) to the base of the power amplifier (Q11). Modulation is applied to both the oscillator and the power amplifier from the transformer (T2) which acts as either the audio output or modulation transformer. The signal is then fed through the tuned circuit (L8, C36) to the antenna change-over switch to the self-contained whip antenna.

The receiver is a crystal-controlled superheterodyne employing only five transistors to obtain a sensitivity of 1 microvolt. The signal passes from the antenna through the tuned circuit (C44, L1) and is amplified in the RF amplifier (Q1). It then goes through the second tuned circuit (L2) for added selectivity to the base of the mixer stage (Q2). The local oscillator (Q3) generates a crystal-controlled signal 455 kHz below the on-channel signal. This signal is coupled from collector of the oscillator through a transformer (L3) to the emitter of the mixer stage (Q2). The resulting signal from this stage is on 455 kHz. This signal is amplified by the 2-stage IF amplifier (Q4, Q5) and then is detected by the diode (CR1). The AVC is developed at the collector of Q5 and fed back to the RF amplifier stage to stabilize the gain and prevent strong signals from overloading the circuits. Additional AVC is taken from the emitter of the RF amplifier (Q1) and coupled back to the base of the first IF amplifier (Q4). The noise limiter circuit follows the detector and then the signal is fed through the squelch circuit (Q6) to the base of the audio driver stage (Q7). The signal is coupled to the push-pull output stage (Q8, Q9) and then through T2 to either the speaker or the transmitter. The audio stages and the antenna are common to both the transmitter and receiver sections of the Messenger.

THE RUSTLER

The Rustler is typical of the older tube-type CB transceivers. There are seven tubes used in this unit which may be operated from either a 12v DC source or from the 117v AC lines. There are seven channels supplied in the Rustler.

The transmitter consists of an oscillator-doubler stage (V1A) which feeds the RF amplifier (V1B). The amplifier is plate modulated from the audio transformer (T3). The transmitter is keyed on by the microphone switch that completes the ground circuit on the oscillator and amplifier cathodes.

The incoming on-channel signal is fed to the grid of the RF amplifier (V2A) and then coupled through capacitor C18 to the grid and plate of oscillator stage (V3B) and fed into the screen grid of the mixer stage. The resulting 455-kHz signal is amplified in the IF stage (V4) and is detected in one section of V5. The other sections of this tube act as the first audio noise limiter and supply the AVC circuits. The audio amplifier-modulator consists of microphone preamplifier (V6B), driver (V6A) and the audio output amplifier (V7).

This unit contains the older type vibrator power supply used with the tube-type equipment. When operating from a 12v DC source, the vibrator changes the 12 volts to a form of DC which is fed to a small section of T4. This steps up the voltage and then the diodes (CR1 and CR2) rectify the higher voltage to supply the B+. When -17 volts AC is used, the vibrator is bypassed and the voltage is fed directly to the lower winding of T4. Again the transformer steps up the voltage and supplies the rectifier circuit.

SBE SIDEBANDER

This CB transceiver has the added capability of being able to transmit and receive single sideband signals. It employs a frequency synthesizer circuit which uses only 14 crystals to obtain the full 23 channels on both transmit and receive. Additional crystals are used in the sideband mode of operation to ensure correct operation on either the upper or lower sideband. When using single-sideband (SSB) operation, more power is available for transmission and the bandwidth is much narrower so better reception is possible.

It must be remembered that in order to receive SSB signals, the unit must be switched to the proper mode. Most of the tests and tuning may be carried out in the AM mode since there is no RF power measureable unless the transmitter is modulated. This is explained in an earlier section of this book.

SSB operation requires a "clarifier" control which is basically a fine tuning control. The correct reception of a signal depends upon the accuracy of both transmitting and receiving frequencies. If the receiver frequency is higher than the transmitter frequency, the voice will be high pitched and if the receiver is lower in frequency the voice pitch will be too low. Therefore, the fine tuning control is necessary. If the incorrect side band is chosen, then no amount of frequency change will make the signal intelligible.

SONAR MODEL T-2

The Sonar unit is a 2-channel hand-held unit that has a power output of one watt. It is quite typical of the lower power portable CB transceivers. There are 14 transistors and six semiconductors used in the T-2. (See the schematic on foldout Panel L.) The transmitter section uses three transistors to develop 1 watt. The oscillator (TR12) has the crystal in the base circuit, and the channel selector switch completes the circuit to ground for the desired crystal. The RF signal is coupled through transformer T4 to the base of the driver stage (RT13) where the power is increased sufficiently to drive the final amplifier. The signal is coupled through L5 to the base of the final stage (TR14). Modulation is applied to both the driver and final amplifier stages from audio transformer T5. The modulated RF signal is then fed from the emitter of TR14 through the pi network (C55, L7, C56) to the antenna through switch SW1-1 and a filter (C4, L3, C3, L2, C2). All three transmitter transistors operate in grounded collector circuits.

The incoming on-channel signal is fed through switch SW1-1 to transformer L8 and then to the emitter of the RF amplifier (TR1). The output of the amplifier is fed through L9 to the base of the mixer stage (TR2). The crystal-controlled local oscillator (TR3) produces a signal 455 kHz below the incoming signal and it is mixed with the on-channel signal in the mixer. The resulting signal is fed through the two IF amplifiers TR4 and TR5 and then is detected by the diode (RD1). The audio signal then goes through the noise limiter (RD2) circuit to the audio amplifier stages.

An adjustable squelch is provided by the diode (RD3) and amplifiers TR6 and TR7. The squelch may be opened at any value from 1 microvolt to 30 microvolts by adjusting R19.

The audio amplifier-modulator consists of four transistors, preamplifier TR8, driver TR9 and the push-pull output stage (TR10, TR11) which develops 120 milliwatts of output to the speaker or to modulate the transmitter.

TELSAT 924

The Telsat 924 (foldout Panel M) is one of the newest types of CB transceivers. It has many of the newer features to be found in the more complex equipment available at the present time. This is an all-transistor unit employing a frequency synthesizer to provide 23 crystal-controlled transmit and receive channels. In addition, it has a separate receiver section so that the emergency channel (9) may be monitored at all times.

There are several features that should be mentioned concerning this unit. Because of the synthesizer circuit, the unit is capable of 23-channel operation with fewer crystals. The transmitter section uses five transistors to obtain an output of more than 3 watts and modulation is applied to both the driver and final amplifier. A highly selective tuned circuit is used in the mixer stage where the transmitter oscillator and synthesizer frequencies are mixed to ensure that no harmonics are allowed to be present in the final stages of the transmitter.

The receiver is a dual-conversion superheterodyne that employs a mechanical filter for added adjacent-channel rejection. The receiver local oscillator also includes a "delta" tuning switch that permits fine tuning of plus or minus 1.5 kHz. This allows better reception from transmitters that may be slightly off frequency. An output has been provided prior to the final audio section to allow tape recording of all communications, both transmit and receive, on the unit.

The monitor receiver section is completely independent of the receiver section except for the audio amplifier portion. This section is crystal controlled and its operation is indicated by a monitor light on the front panel of the transceiver. This light is activated by the squelch circuit in TR7. When an incoming signal on Channel 9 overcomes the squelch, the amplifier circuit (TR8 and TR9) is activated, causing the monitor light to flash. The switch (SW-4) is a pushbutton type which transfers the audio from the standard receiver detector output to the output of the monitor section and thus allows the message on Channel 9 to be heard.

Index

S

T

Z